12,000 BC

12,000 BC

TWO EXTINCTION-LEVEL EVENTS WE KNOW NOTHING ABOUT

Jerry Kroth, PhD

Algora Publishing
New York

Library of Congress Cataloging-in-Publication Data

Names: Kroth, Jerome A., author.
Title: 12,000 BC : two extinction-level events we know nothing about / Jerry Kroth, PhD.
Other titles: 12,000 BCE
Description: New York : Algora Publishing, [2024] | Includes bibliographical references. | Summary: "Jerry Kroth, PhD, explores two recently discovered apocalyptic events that shaped this planet and human cultural evolution. A gigantic cometary event impact plus as a solar storm, called a Miyake event, disrupted the ice age. Some 35 species were suddenly rendered extinct; the human race almost disappeared as well"— Provided by publisher.
Identifiers: LCCN 2024052226 (print) | LCCN 2024052227 (ebook) | ISBN 9781628945546 (trade paperback) | ISBN 9781628945553 (hardcover) | ISBN 9781628945560 (pdf)
Subjects: LCSH: Mass extinctions. | Antiquities, Prehistoric. | Prehistoric peoples. | Human-alien encounters. | Extraterrestrial anthropology.
Classification: LCC QE721.2.E97 K76 2024 (print) | LCC QE721.2.E97 (ebook) | DDC 576.8/4—dc23/eng/20241108
LC record available at https://lccn.loc.gov/2024052226
LC ebook record available at https://lccn.loc.gov/2024052227

Cover image: Youtube channel "Destiny Space."
Public Domain, https://www.youtube.com/@DestinySpace

Printed in the United States

For David Griffard

Table of Contents

PART ONE: 12,000 BC 1

Deep-Rooted Misconceptions about Early Civilizations (or man?) 1

Disclaimers 5

Four theories 7

Theory #1: The academic consensus: Human hunter-gatherers did it. 7

Contrary evidence 8

Theory #2: A comet hit Earth causing the Younger Dryas and the megafauna extinction 10

The evidence 12

Platinum in South Africa 16

Fires and biomass burning 16

Theory #3: Not a comet but a supernova! 18

Theory #4: It wasn't a comet, nor a supernova, it was a solar flare that caused it all 21

Could it have been two catastrophic events, not one? 27

More on the Clovis people 32

Other evidence of massive cataclysms: bone-beds 35

Texas and Nebraska 36

Conclusion 37

PART TWO. HUMANITY THROUGHOUT THE YOUNGER DRYAS 41
Myth, Memory, and Mnemonics 41
Myths which lead to facts 44
What about Atlantis? 45
Some factoids about Atlantis mythology 47
The location of Atlantis 49
An absence of archeological discoveries 51
A strange coincidence 53
Conclusion 53
More circumstantial evidence: The lost city of Cuba 56
Location of 'lost city' 57
Hapgood's Crustal displacement 59
An Update on Oronteus Fineas 62
Gobekli Tepi 66
The Yonaguni Monument, Japan 70
The Jomon culture 74
The pyramid of Gunung Padang 75
Summing up 78

PART THREE. A QUIETLY REBELLIOUS THEORY 83
Why I am telling you this? 84
Pictographs and pyramids: an archeological journey 85
Examples with a higher level of authenticity 90
India cave painting 101
Chinese, Mexican & Egyptian Pyramids 105
Pyramid planes of Giza (left) and China (rt) 109
Dolichocephaloids 117
A trip to Malta 121
A different approach to interpretation: John Frum and psychoanalysis 122
Pi and phi 126
Tough talk from the mainstream 139
The Voynich Manuscript 140
Summary: Archeology unexamined 144

EVOLUTIONARY BIOLOGY AND THE EXTRATERRESTRIAL 149
Kirlian photography and ch'i. 150
Kirlian fingerprint photo 150
Man as hybrid 151
Denisovans 155

How old are we? 156
Mungo Man, Chimps, and H. sapiens Genes 158
Traditional explanations 165
The development of agriculture: 165
The frontal lobes 166
The mythological theory expressed 168
More on the efflorescence of human culture 178
Connecting the dots 180
A new twist is added in 2020. . . 189
Curious new genes 190
Back to psychoanalysis 195
CONCLUSION AND SYNTHESIS 203
Academic political correctness 206
ABOUT THE AUTHOR 209
PICTURE CREDITS 211

PART ONE: 12,000 BC

Deep-Rooted Misconceptions about Early Civilizations (or man?)

> When we were in high school, most of us learned that Homo sapiens was around for 50,000 years. That was a mistake. A skull of an anatomically modern human was found in South Africa dating to 196,000 years; we call it the Omo skull.[1]

That first misconception means we start with a large dose of humility — we know less of our history than we think, and modern science has a lot to teach. Indeed, human beings, *Homo sapiens*, have been here for close to 200,000 years!

But this book is really investigating a different issue, and it relates to whether or not human culture and civilization started up with the end of the ice age. The ice melted, the ocean rose, and temperatures warmed allowing agriculture to take hold. With agriculture came settlements, and humans were no longer scattered, nomad hunter-gatherers, but could be living in larger groups in specific areas. Not everyone was engaged in food production; written language, tool making, and culture also got a big boost.

The ice age formally ended about 10,000 years ago. The era before that is called the Pleistocene; after that we talk about the Holocene.

You can see from this graph that ice core samples have revealed how the temperature started to climb about 10-15,000 years ago. The Earth warmed, and human civilization flowered.

---

[1] The University of Utah, "The Oldest Homo Sapiens," UN News Archive, March 25, 2011, https://archive.unews.utah.edu/news_releases/the-oldest-homo-sapiens/

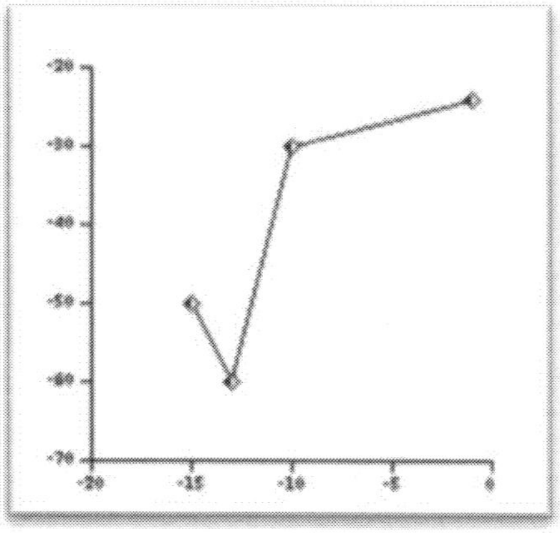

Greenland ice core temperatures

Why, you ask, is this a misconception? Well, that will take some explanation, so let us begin in a roundabout way. Let us start with something that happened, something quite big, catastrophic really, which we knew and still know very little about it.

First, from about 12,000 BC (14,000 years ago) to about 10,000 BC, a great number of animals went extinct. It wasn't just that the Earth was warming. Something else was going on. The Woolly Mammoth roamed this planet for 300,000 years and then suddenly disappeared.

The Saber Toothed tiger, enjoying life on this planet for 56 million years—*and weathering many ice ages*—also suddenly went extinct.

He (or she, or they) survived all the ice ages you see in the figure below. So we have to ask, what exactly happened so that they suddenly vanished in the last 12,000 years?

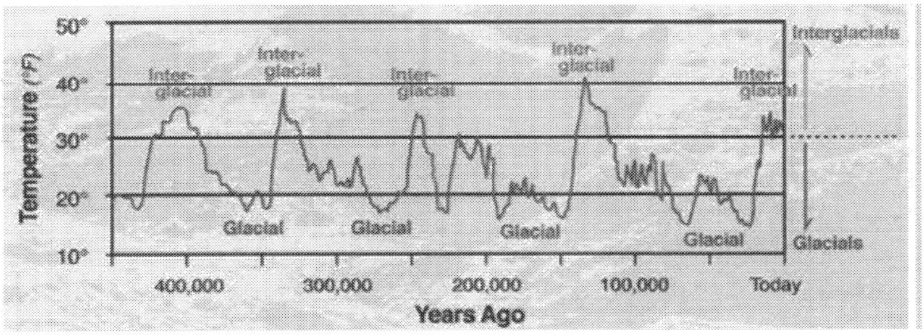

Glacial-Interglacial Cycles of the past 450,000 years

In fact, 35 different species went extinct between ,000 to 10,000 years ago. Take the giant sloth as an example. Sloths roamed this planet for millions of years, but perished in this same, geologically "short" period. [2]

Here is a composite picture of some of the animals that disappeared in this period of time.

[2] James Ashworth, "Giant Sloth that roamed South America scavenged for meat," Oct. 2021, Natural Museum of History. https://www.nhm.ac.uk/discover/news/2021/october/giant-sloth-that-once-roamed-south-america-scavenged-for-meat.html

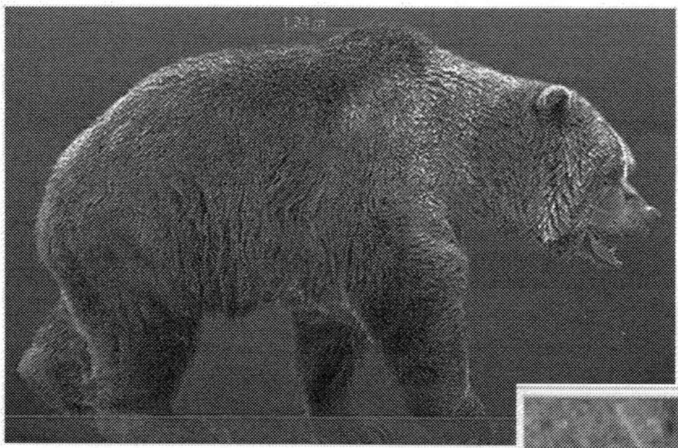

Cave bear (*Ursus spelaeus*) reconstruction

The dhole (*Cuon alpinus*), now restricted to the southern portions of Asia, was present from Iberia to Mexico during the Late Pleistocene

Reconstruction of *Mixotoxodon larenis*, a toxodontid notoungulate

But it just wasn't in the Americas that this happened. Many died off in Asia, like the Formosan leopard in Australia, and all over the world.

So, yes, something big happened in the interval of geological time from 14,000 to 10,000 years ago—or to use more biblical dates, 12,000 BC to 8000 BC.

Disclaimers

Before we go much further, I want to remind you that my Ph.D. is in psychology, not anthropology, archeology, solar physics, archeo-astronomy, or paleontology, so this is all outside my wheelhouse. My professional interests are in the psychology of denial, suppression, propaganda, and delusions; and, yes, there is plenty of that on this topic, but we will get to that later. For the present, my role is to be merely a scientific journalist and report to you the science as best I can.

Moreover, this is not going to be "fringe" or "conspiracy theory" at all. It is generally coming straight from peer-reviewed journals. And it isn't antiquated literature, either. In 2024 the *New York Times* published a piece looking at whether a comet hit Earth and caused all this chaos.[3]

Again, we are not delving into conspiracy literature. We are looking at a deeply perplexing scientific problem: what happened 14,000 years ago that turned everything upside down. And we will be looking at very new scientific data, almost all of it coming at us in the last 15 years.

[3] Zack St. George, The Comet Strike Theory that just won't die." New York Times, 5 March, 2024https://www.nytimes.com/2024/03/05/magazine/younger-dryas-impact-hypothesis-comet.html

Four theories

How do scientists explain the unusual extinction of so many animals? Actually, no one knows for sure. There are four, maybe five, major theories to explore.

When we talk about mammals like the Woolly Mammoth and the Saber Toothed tiger, "82% of these animals disappeared in North America, 74% in South America, 71% in Australasia, 59% in Europe, 52% in Asia, and only 16% in Sub-Saharan Africa."[4]

Those are the facts we are hoping to explain with a review of how all that might have happened.

Theory #1: The academic consensus: Human hunter-gatherers did it.

In the mainstream view, human hunters killed off all the big animals for food. As humans entered Australia, the author of *Guns, Germs and Steel*, Jared Diamond, opines all the big animals succumbed to hunting. Ditto for the entrance of humans into the Americas 13-15,000 years ago and for Eurasia as well. They were hunting in packs with spears. In a peer-reviewed article, the authors write:

> "The end of the Pleistocene was characterized by an intense, highly selective extinction event, affecting large-bodied terrestrial mammals worldwide. This period coincides with two major events, the last glacial period, and the dispersal of technologically advanced *Homo sapiens* outside the Old World. Upper Paleolithic hunters were armed

[4] Richard Firestone "Disappearance of Ice Age Megafauna," Planetary Science, July 24, 2021, https://beta.capeia.com/planetary-science/2019/06/03/disappearance-of-ice-age-megafauna-and-the-younger-dryas-impact

> with advanced, projectile weapons, and were able to dispatch prey as large as mammoths." [5]

In other words, most of these extinctions fell at the feet of human hunters.

Contrary evidence

That sounds plausible but there are some issues. For example, *Homo Sapiens* was in China 68,000 years ago, in Australia 60,000 years ago, in Europe 56,000 years ago. Most of the extinctions happened between 14,000 and 10,000 years ago. Why did it take the hunters in Australia 40,000 years to hunt down all the big animals but only a couple thousand years in the Americas?

To put that in a different light: if human hunters are responsible for the extinction of mammals 10–14,000 years ago, it only took these hunters 2–3000 years to kill off 80% of the mammals in the Americas—because humans had only just arrived 2–3000 years earlier. But if we follow out that logic, it took 31,000 years for human hunters to kill 59% of the mammals in Australasia—because they were there for at least 45,000 years.

These extinctions took place 10-14000 years ago, but human hunters were in China, Australia, and Europe *long before* these extinctions occurred in these locations. That doesn't ring true, despite its currency in the academic world.

We can't gaslight *Homo sapiens* for all of the extinctions. The timing is off. Don't forget, humans were in Africa for 196,000 years and Africa suffered *the least* number of megafauna extinctions, only 16%.[6]

One way out of this dilemma is that recent research suggests that humans came to the Americas earlier, dating to 23,000 years. That would imply they had more time to kill off all the megafauna in North and South America. [7]

In Northwest Europe human hunters are dated at 45,000 years, which means they had 35,000 years to kill off all the big animals.[8]

So we don't get confused, let's put this in a table.

---

[5] F. Coratunuta, et. al., The Well Behaved Killer, Paleogeography, Palaeoclimatology, Paleoecology, Vol 500, 1 July 2018, Elsevier, https://www.sciencedirect.com/science/article/abs/pii/S0031018218300725

[6] Hannah Ritchie (2022) - "Did humans cause the Quaternary megafauna extinction?" Published online at OurWorldInData.org. Retrieved from: 'https://ourworldindata.org/quaternary-megafauna-extinction'

[7] Thomas Kissell, Ice Age Footprints Rewrite History of Humans in America, 22, Aug, 2023, Greek Reporter,https://greekreporter.com/2023/08/22/ice-age-footprints-white-sands/

[8] J. Hublin, et. al., *Homo sapiens* already reached Northwest Europe 45,000 years ago, Max Plank Institute 31 Jan. 2024. https://www.mpg.de/21451099/0125-evan-homo-sapiens-reached-northwest-europe-more-than-45-000-years-ago-150495-x

Table 1. Die-offs across Earth

Continent or country	Years humans in place	Number of years to kill off all mega-fauna by 10,000 BC	Percent of megafauna going extinct
Northwest Europe	45,000	35,000	52%
Americas	23,000	13,000	82%
Australasia	60,000	50,000	59%
China	68,000	58,000	NA
Africa	196,000	186,000	16%

*The data on percent that went extinct in China is unavailable.[9]

If megafauna extinctions occurred 10–14,000 years ago blaming *all* the extinctions on human hunter-gatherers now appears as a wild speculation, despite the fact that it is the au courant theory in anthropology today. Why didn't humans hunt them to extinction in Africa? They didn't. Humans were there for 196,000 years, but only 16% of the large animals went extinct in Africa. But 82% in North America? These are incredible discrepancies.

If we grant newer data and recent research that humans arrived in the Americas over 23,000 years ago, then would have taken 13,000 years to hunt these animals to extinction but that is still *three times as long* in Europe? The Americas are four times larger than Europe!

Let me express this contradiction another way to make the absurdity even clearer: The Americas are *four times larger than Europe*, yet humans hunted down all the megafauna *three times faster* in the Americas than they did in much smaller Europe.

This popular theory just doesn't compute.

One final problem is that the people who lived in North America at that time were called a specific branch we call the Clovis people. If they slaughtered all the big animals, they seem to have gone right out with them. "The indigenous native people in America (Clovis culture) disappeared...and it coincides directly with the extinction of North American animals."[10]

[9] See Turvey, S., et. al, Holocene survival of Late Plistocene megafauna in Chaina, Quaternary Science Reviews 2013

[10] Erik Lovin, How did the Younger Dryas touch our lives, 21 May, 2023, Science in your Life https://www.linkedin.com/pulse/how-did-younger-dryas-event-touch-our-lives-erik-lovin/

Also, if hunter-butchers were driving all these animals into extinction, where is the evidence for the butchering? There is some literature on that, but very little: "Critics of overkill have emphasized the small number of securely documented Clovis kill sites considering the magnitude of the slaughter implied."[11]

That represents a rationale of why we might be skeptical that all these megafauna extinctions happened *because* human hunters killed them all. The human overkill hypothesis doesn't really pass the smell test.

Theory #2: A comet hit Earth causing the Younger Dryas and the megafauna extinction

Graham Hancock, whom mainstream archeologists do refer to as "fringe," created an intensely interesting video entitled *Ancient Apocalypse* where he outlines the evidence that a comet struck Earth, likely in fragments, and he posits that this is what caused all sorts of chaos and may be responsible for these mass extinctions. It happened in the Northern Hemisphere—which is why Africa was spared.

As we proceed with this theory, let me introduce the Younger Dryas. This is a term for a peculiar blip in Earth's temperature. As the Earth was starting to warm at the ending of the Ice Age, something peculiar happened, and temperatures plunged back down into the depths of cold for another thousand years. Then, once again, the Earth's temperature returned to warming until we ended up with the climate we have today. Here is a graph of the Younger Dryas.

The data is from Greenland. You can see that 15,000 years ago the Earth started warming—*abruptly!*— and then 14,000 years ago, it plunged back, close to 20 degrees colder. Finally, 12,000 years ago it abruptly warmed again, up until the present day.

[11] Jack Broughton and Elic Witzel, Population reconstructions for humans and megafauna suggest mixed causes for North American Pleistocene extinctions, 21 Dec, 2018, Public Med Central, https://www.ncbi.nlm.nih.gov/pmc/articles/PMC6303330/

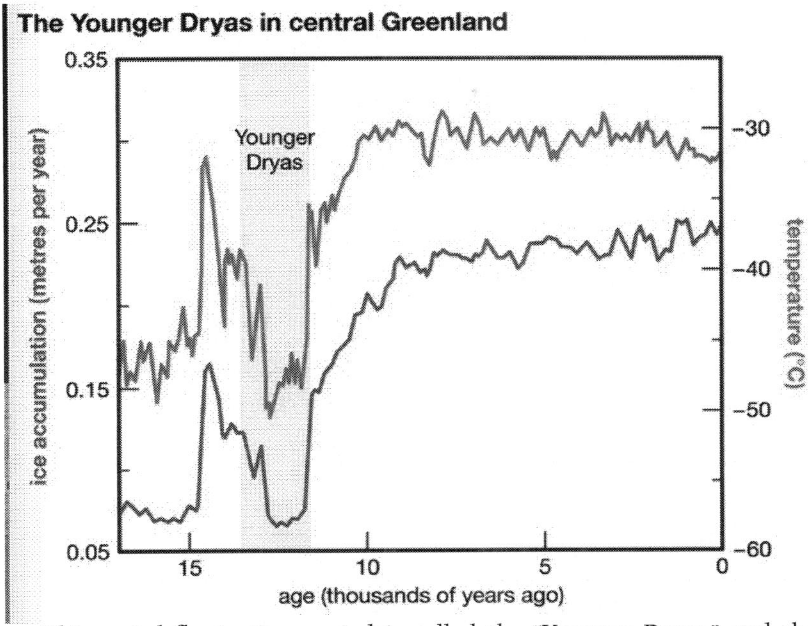

This weird fluctuating period is called the "Younger Dryas," and the period just before it, from 14,570 years to 12,000 years, goes by the name of the Bolling-Allerod Interstadial.

What caused these fluctuations? This theory postulates a comet hit the Northern hemisphere. More than a few academics believe the quick change in temperature may have caused some of these extinctions. [12] But, after all, those animals were living quite well throughout the maximum ice age and, in fact, through many ice ages. Why would a sudden return to cold conditions cause them to perish?

Hancock and others have more detailed explanations and speculate that the comet hitting Earth was the culprit not only for the Younger Dryas but for the extinctions as well.

Here is their theory in brief:

> They had uncovered evidence, they said, that 12,900 years ago, a comet — or possibly a whole fleet of comets — struck Earth and changed the course of history. For the preceding two and a half million years, through the Pleistocene Epoch, the planet's climate fluctuated between frozen stretches, called glacials, and warm interglacials. At that time, Earth was warming again, and the ice sheets

[12] https://www.su.se/department-of-geological-sciences/about-the-department/about-us/tracing-climate-shifts-using-sediment-dna-from-hall-s-cave-texas-1.578012

that covered much of North America, Europe, and Asia were in retreat. Mammoths, steppe bison, wild horses, and other enormous mammals still wandered the Americas, pursued by bands of humans wielding spears with fluted stone blades. Suddenly, somewhere over the Upper Midwest — an explosion.[13]

The evidence

There is a lot of sound research, recent research, that supports their hypothesis. From the University of South Carolina Professor Andrew Moore extracted sediment samples which were dated to the beginning of the Younger Drays. They contained a large amount of platinum and soot. The soot indicates huge wildfires occurred, while the platinum points to a cometary collision. (Platinum generally comes from extraterrestrial sources.)[14] This peer-reviewed piece, authored by four scientists, proposes that the comet caused a very high-temperature impact.

Measurements were taken on three continents, and all the three sites showed microspherules: "siliceous scoria-like objects, corundum, mullite and other rare meteoritc minerals that only form under high temperatures."[15] They also found nanodiamonds and meltglass "indicating enough heat and pressure was present to fuse materials on the Earth's surface." [16]

A comet impact would explain such high temperature impacts and fused minerals. Researchers may be at a loss to explain why or how the extinctions happened, but they are quite certain a comet or a series of cometary fragments crashed into Earth 12,900 years ago and set off a barrage of changes, culminating in the Younger Dryas.

The theory alleges the impact caused massive wildfires where soot blocked the sunlight and brought on the cold spell of the Younger Dryas.

Their article was published in the *Proceedings of the National Academic of Science.*

[13] Anna Lindholm, Tracing climates shifts using sediment DNA from Hall's Cave, 14 Oc 2021, Stockholm University https://www.nytimes.com/2024/03/05/magazine/younger-dryas-impact-hypothesis-comet.html

[14] Dominique Tonner, "The platinum-group metals in your car come from ancient meteroites that bombarded young Earth, ABC Science. 17 Aug 2019.https://www.abc.net.au/news/science/2019-08-18/platinum-group-metals-car-exhaust-ancient-meteorites/11417812

[15] Ted E. Bunch, et., al Very high-temperature melt products as evidence for impacts 12,900 years ago, 18 June 2012 PNAS https://www.pnas.org/doi/full/10.1073/pnas.1204453109

[16] Carol Ward, U of SC archeologists finds evidence for extinction theory, 12, Oct 2019, University of South Carolina https://sc.edu/uofsc/posts/2019/10/10_chris_moore_research.php

Richard Firestone, a nuclear chemist at the Lawrence Berkeley National Laboratory, pulls us deeper into the comet impact theory. He discovered a layer of material called the "black mat" which can be seen as the black line in the accompanying photo. That indicates fire, soot, and smoke played a role in mass extinctions. His analysis says the black mat layer is composed of microspherules:

It appears that these microspherules emerged from a cloud of molten iron droplets, colliding with each other as they cascaded to Earth. The microspherules are geochemically and morphologically comparable with cosmic impact material from Meteor Crater, Arizona, and with material from the Trinity nuclear airburst in Socorro, New Mexico, and inconsistent with anthropogenic, volcanic, and authigenic materials ....Beneath the black mat and at all YDB (Younger Dryas) sites is found charcoal, vesicular carbon spherules, glassy carbon, aciniform carbon (AC/soot), and nanodiamonds . . .which are all evidence of associated burning... . Profound evidence of widespread burning at the onset of the YD is also seen in the Greenland ice cores where a massive peak in ammonium (NH4), a marker biomass burning, is observed at 12,822 ± 1 cal yr BP. Other markers including nitrate, oxalate, acetate, and formate, indicate that this was the largest biomass burning episode in at least 120,000 years. CO2 evidence from Antarctic ice cores suggests that fires may have consumed ≈ 9% of Earth's biomass.[17]

Nanodiamonds, or diamond nanoparticles, are diamonds with a size below 100 nanometers. They can be produced by impact events such as an explosion or meteoritic impacts. Source: https://en.wikipedia.org/wiki/Nanodiamond

Putting Firestone's work more closely into layman's terms, his research implies a comet struck, followed by a planetary wide fire that burned up *nine percent of the Earth's vegetation* leaving this telltale black mat of evidence. This event, in turn, is what contributed to the extinctions across the planet

[17] David Biello Did a comet hit Earth 12,000 years ago, Scientific American, Jan 2, 2009

consuming 82% of animals in North America, and most of these extinctions were rapid and spontaneous. [18]

Firestone did not identify the actual comet or how many fragments hit, nor any specific craters. There is one crater in Greenland, known as the Hiawatha crater, almost 31 kilometers wide, which may be a candidate. After the fires, a layer of soot and dust blocked the sun, and the impact also affected the ozone layer, initiated the cold spell of the Younger Dryas and had long-lasting effects on photosynthesis and thus a prolonged extinction process.

> In either case, the blast from this impact would devastate an area up to 1,300 km from the impact...This dust would lead to a sudden global cooling, worldwide drought, and failure of photosynthesis, for an extended period of time...Substantial NOx and SO2 would be produced by this impact, enough to destroy Earth's ozone layer exposing all life to increased ultraviolet radiation and increased acid rain . . .[19]

Firestone has substantial data to support the theory, but it is controversial to more staid archeologists of today.

> The evidence for the YD (Younger Dryas) impact is compelling but continues to receive widespread criticism from a small number of purists. Attempts by reviewers to suppress the publication of the data have generally failed, and arguments against the data-based conclusions are largely anecdotal.[20]

This is not attributable to Firestone alone, however; another scholar, astronomer Bill Napier, similarly argues for it:

> The Younger Dryas Boundary, which is marked by catastrophic wildfires over North America, the extinctions of 35 genera of mammals, and widespread soot, "magnetic grains and nanodiamonds" was caused by the impact of a dense swarm of debris from a disintegrating large 50-100 km short period comet with Earth.[21]

[18] Late Pleistocene Extinctions, Wikipediahttps://en.wikipedia.org/wiki/Late_Pleistocene_extinctions

[19] Ross J. Salawitch (Lead Author), Laura A. McBride, Chelsea R. Thompson, Eric L. Fleming, Richard L. McKenzie, Karen H. Rosenlof, Sarah J. Doherty, David W. Fahey, *Twenty Questions and Answers About the Ozone Layer: 2022 Update, Scientific Assessment of Ozone Depletion: 2022*, 75 pp., World Meteorological Organization, Geneva, Switzerland, 2023. https://csl.noaa.gov/assessments/ozone/2022/twentyquestions/

[20] Richard Firestone, Disappearance of ice age megafauna 24 Jul 2019, Capeia https://beta.capeia.com/planetary-science/2019/06/03/disappearance-of-ice-age-megafauna-and-the-younger-dryas-impact

[21] NASA's Hearsarc, https://heasarc.gsfc.nasa.gov/docs/heasarc/headates/earlier.html

Platinum in South Africa

One study from South Africa shows unusual deposits accompanied by curious carbon dating further supporting the impact hypothesis:

> The team, led by Professor Francis Thackeray of the Evolutionary Studies Institute at the University of the Witwatersrand in Johannesburg, South Africa, discovered evidence of a remarkable "platinum spike" at a site called Wonderkrater in the Limpopo Province, north of Pretoria in South Africa...Thackeray discovered this evidence from a core drilled in a peat deposit, notably in a sample about 12 800 years old. This research was published in *Palaeontologia Africana*.
>
> .... Thackeray said, "Our finding at least partially supports the highly controversial Younger Dryas Impact Hypothesis (YDIH). We seriously need to explore the view that an asteroid impact somewhere on Earth may have caused climate change on a global scale, and contributed to some extent to the process of extinctions of large animals at the end of the Pleistocene, after the last ice age.[22]

Fires and biomass burning

One additional article puts the nail in the coffin. It is a peer-reviewed article authored by 20 scientists arguing that there were severe fires (biomass burning) and these deposits were found in ice cores dating to 12,800 years, *almost the exact time of the comet impact.* (emphasis mine).[23]

That pretty much sums up the evidence that a comet impacted Earth 12,900 years ago! As one publication put it:

> There is little doubt that a megaflood of glacial meltwater cascading off the North American continent into the Atlantic Ocean spurred the birth of agriculture and civilisation in the Middle East around 12,900 years ago. What was not known until recently is that this event, known as the Agassiz megaflood, may have been triggered by a comet exploding above or plunging into the ice sheet north of the modern Great Lakes. According to two geologists at the University of Oregon, Dr. Douglas Kennett and Dr. Jon Erlandson, there is reason to believe a large chunk of a comet exploded above or crashed directly into the Laurentide ice sheet, rupturing the ice dams on the easterly margin of Lake Agassiz and causing frigid water to flood into

[22] Keith Cowing, African Evidence supports Younger Dryas Impact Hypothesis 3 Oct 2019, Space Ref https://spaceref.com/science-and-exploration/african-evidence-supports-younger-dryas-impact-hypothesis/

[23] Wendy Wolbach, et. al., "Extraordinary Biomass-Burning Episode and Impact Winter Triggered by the Younger Dryas Cosmic Impact 12,800 Years Ago." 2. Lake, Marine, and Terrestrial Sedimentshttps, 2018 doi/abs/10.1086/695704

the North Atlantic. The primary evidence consists of a carbon-rich layer of soil, dating to around 13,000 years ago, found at 50 Clovis-age sites across North America. The Clovis people, the first wave of human colonists to reach North America from Siberia, had only recently colonized the Americas.[24]

The Willamette Meteorite. Source: https://en.wikipedia.org/wiki/Willamette_Meteorite#/media/File:At_American_Museum_of_Natural_History_2024_005.jpg

Further,

The comet, it was said, fragmented into a number of pieces as it hit in a cascade over the Northern Hemisphere. Some of the fragments actually were carried by flood waters. The 15-ton Willamette Meterorite, found in Oregon, is thought to be a remnant of that cosmic cataclysm and carried south by the flood. It is also the largest meteorite

[24] Peter Belwood, Farmageddon 19 Aug, 2007 ,The Age, https://www.theage.com.au/technology/farmageddon-20070819-ge5mar.html

> ever found in the United States (see photo). It is appropriately dated at roughly 13,000 ybp.[25]

That sums up theory #2 about cometary impact and the Younger Dryas extinctions. It is, in my view, very compelling.

But before Dr. Belwood's text went to press, another study by a team led by Prof. James Kennet of UC Santa Barbara from the Department of Earth Sciences confirms a comet hit and that it took a large portion of the human population with it. Furthermore, he puts it at 12,800 years ago as well.[26]

Theory #3: Not a comet but a supernova!

Astronomer Robert Brakenridge in another peer-review piece proposed a supernova arrived at our planet about 12,900 years ago, and the supernova, not a comet, was the culprit. It released large quantities of high-energy photons and X- ray bursts at the beginning o the Younger Dryas.[27]

He believes this event might even have initiated the massive extinctions at that time too. The gamma rays, X-rays, and UV radiation emitted deplete the ozone layer and damage living beings as well as start wildfires.[28]

> In seconds, billons of times the energy of the entire Sun is released; a wave of destruction moves through the doomed star from the inside out, blasting away an octillion tons of gas, sending it raging into space at a significant fraction of the speed of light.... When the core collapsed, creating the explosion that would tear its outer layers away, it compressed into a ball of neutrons, an object with the mass of a star like the Sun but all contained in a sphere just a few kilometers across..[29]

Here's a headline:

[25] "The Willamette Meteorite," Wikipedia. https://en.wikipedia.org/wiki/Willamette_Meteorite

[26] Dave Adalian, "Did a comet burst crash Earth's Climate 12,800 years ago." Earht Sky, 5 July 2024 https://earthsky.org/space/younger-dryas-comet-burst-crashed-earths-climate/

[27] Robert Brakenridge,Core-collapse supernovae and the Younger Dryas/terminal Rancholabrean extinctions,Icarus,Volume 215, Issue 1,

2011.https://www.sciencedirect.com/science/article/abs/pii/S0019103511002612

[28] A Supernova Destroyed Some of Earth's Ozonefor a Few minutes in 2022 New York Times, https://www.nytimes.com/2023/11/14/science/gamma-ray-burst-ozone-depletion.html

[29] Phil Plait A supernova that left chaos in its wake, 24 Apr, 2020,SYFY

https://www.syfy.com/syfy-wire/a-supernova-that-left-chaos-in-its-wake

Earth is hit by blast of energy from a dead star so powerful that scientists can't explain it

After it exploded, the remains or "remnant" is the bluish-red material displayed in this 2022 photo.[30]

Supernova remnant SNR E0519 -69.0 in the Large Magellan Cloud

Brakenridge holds that the explosion hitting Earth had numerous effects, one of which was to deplete the Earth of oxygen, another reason for the extinctions.

As part of his evidence, he cites tree ring data and shows radiocarbon levels jumped by 3%, "a staggering increase" he says.[31]

[30] Keith Cowing, the Ghost of a giant star, Vela Supernova, Space Ref, 13 Nov, 2022. https://spaceref.com/science-and-exploration/the-ghost-of-a-giant-star-vela-supernova-remnant/

[31] Plait, Ibid.,.https://www.syfy.com/syfy-wire/a-supernova-that-left-chaos-in-its-wake

In trying to understand how a supernova could explain all this, I wrote directly to Dr. Brakenridge, and he kindly replied that some of the impacts...

> would have been inozation of ozone (03) and conversion to nitrate, and N01. Ozone destruction causes a major increase in solar UVB light reaching the surface and it is a greenhouse gas, so also a strong cool, Ergo Younger Dryas. Extinctions are plausible; especially for large mammals...he concludes "I think the evidence is actually quite strong that the Vela supernova caused the extinctions and the Younger Dryas cold period.[32]

I tried to find corroboration for his assertions and wrote to Dr. Gunther Korschinek, a physicist from the Technical University of Munich, a specialist on supernovae. He said a supernova and the associated gamma-ray bursts and ultraviolet radiation could perhaps have a significant impact on life on Earth as well as inducing climate cooling,[33] but only if it were close—say, 50 light years from Earth. He believes the Vela supernova was too far away (800 light years) to have caused these effects.

No corroboration for Brakenridge, in other words, from Dr. Korshinek—but let's not give up just yet.

In looking further, there are not that many articles on the Vela supernova *causing* the Younger Dryas and the consequent extinctions, but there are articles in the literature looking at a supernova's role in *prior* mass extinctions on Earth, millions of years earlier. In these pieces, the impact on the ozone layer appears to result in major effects on the plankton and reef communities with disastrous impacts on the food chain. [34]

Author Paul LaViolette, Ph.D., does seem to corroborate Brakenridge's work and believes that it was a supernova burst which was the true culprit. He explains how that celestial event caused the extinctions and "the demise of the megafauna would have been due to a number of factors" which he then lists as follows:

- Death or illness due to an excessive dose of ionizing radiation, both UV and cosmic rays, due to elevated solar flare activity and a ground-contacting solar proton event.
- Death due to famine since vegetation would have withered during the cool arid Younger Dryas climate and preceding Intra Alleröd Cold Peak.

[32] personal email

[33] Handbook of Supernovae, Springer International Publishing, 2020 https://link.springer.com/referencework/10.1007/978-3-319-20794-0

[34] J. Ellis, and DN Schramm, Could a nearby supernova explosion have caused a mass extinction Proc Natl Acad Sci U S A. 1995 Jan 3; 92(1): 235–238.

doi: 10.1073/pnas.92.1.235

- Death due to heat stroke and smoke inhalation from exposure to the firestorm that produced the black mat layer evident today in North America and Europe.
- Death due to drowning in the glacial meltwater deluges that swept across the continent.
- Death due to poisoning by noxious fumes released from the meteoric ice vaporized in the atmosphere."

It is hard for a layperson like myself to get a handle on the dynamics of this catastrophe. What comes first? Fires, floods, UV radiation, freezing, smoke, oxygen deprivation and suffocation, ozone depletion or whatever effluvium follows? It is a gnarly task to tease out exactly how the cataclysm unfolded.

When we realize we are trying to figure out how a sequence of events unfolded 12,900 years ago, we have to admit we are speculating rather wildly. But for the present, let us just conclude that this third theory, a supernova explosion hitting Earth, has some impressive evidence supporting it.

Theory #4: It wasn't a comet, nor a supernova, it was a solar flare that caused it all

A final theory asserts it was a cosmic event, but one involving our own sun.

A consortium of scientists from the University of Leeds discovered in 2023 that there had been a huge spike in radiocarbon levels 14,300 years ago. The spike they believe was caused by a massive solar storm, the biggest ever identified, far bigger than the 1859 Carrington Event.

They matched tree rings with Greenland ice cores and found their data corresponded to measurements of beryllium in Greenland. "The team proposes that the spike was caused by a massive solar storm that would have ejected huge volumes of energetic particles into Earth's atmosphere." [35]

This was also a peer-reviewed study. A second analysis drew the same conclusion that a massive solar event occurred at that time:

> A SPE of this magnitude would have been large enough to deliver a lethal radiation dose of at least 3 – 6 Sv to the Earth's surface, and hence could have been a principal cause of the final termination of the Pleistocene megafauna....The event time-correlates with a large magnitude acidity spike found at 1708.65 m in the GISP2 Greenland ice record, which is associated with high NO-3 ion concentrations and a rapid rise in 10Be deposition rate [beryllium], all of which are indicators of a sudden cosmic ray influx. The depletion of nitrate ion within

[35] Ibid.

this acidic ice layer suggests that the snowpack surface at that time was exposed to intense UV for a prolonged period which is consistent with a temporary destruction of the polar ozone layer by solar cosmic rays. The acidity event also coincides with a large magnitude, abrupt climatic excursion and is associated with elevated ammonium ion concentrations, an indicator of global fires.[36]

Preserved tree-rings in the subfossil stumps cover the past 15 000 years. CÉCILE MIRAMONT

Fossilized tree rings provided the basis for their conclusions.

This data corresponds closely to the Supernova data, but here he argues that a massive solar storm produced these effects, not a supernova. "The radiocarbon and cosmogenic beryllium evidence examined here strengthens suggests that the mass extinction *had a solar cause*. Several studies indicate that toward the end of the ice age the Sun was far more active than it is today."[37]

Even though the 1859 Carrington Event is the largest solar storm in modern history, this event is in another category altogether and is called a *Miyake event*, many times larger than Carrington and easily able to cause signifìcant damage

In a study released Monday researchers identified what appears to be the largest solar storm to hit Earth, estimated to be larger than

[36]Researchers identify largest ever solar storm in ancient 14,300 year old tree rings, October 9, 2023, University of Leeds, https://www.cambridge.org/core/journals/radiocarbon/article/evidence-for-a-solar-flare-cause-of-the-pleistocene-mass-extinction/799A4720B75245B4627AFE412661B3B6

[37]Paul La Violette, "Evidence for a Solar Flare Cause of the Pleistocene Mass Extinction," Radiocarbon 53, pp. 303-323, Jan 2011, https://www.researchgate.net/publication/305971138_Evidence_for_a_Solar_Flare_Cause_of_the_Pleistocene_Mass_Extinction

the Carrington Event by an order of magnitude. The storm occurred 14,300 years ago, but is evidence of a yet unknown dimension of the sun's extreme behavior and hazards to Earth.[38]

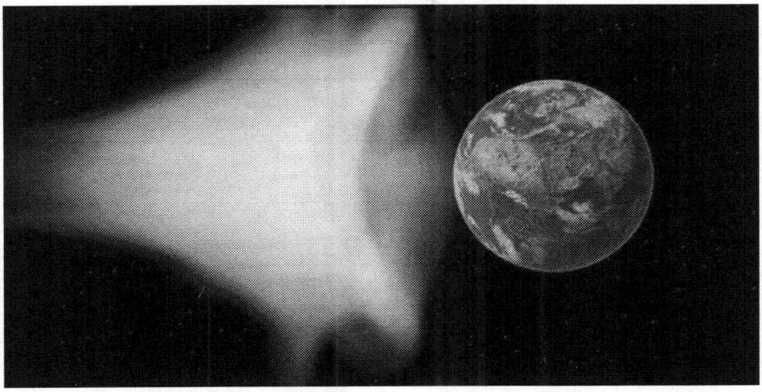

Miyaki solar storm

> We have no way to compare how large a storm like this was with anything in recorded history, says Heaton. The largest solar storm we have evidence for, the so-called Carrington event of 1859, sparked fires and induced currents in telegraph wires, but this specific event was so small compared with Miyake events that it wouldn't even register a blip in the radiocarbon record.[39]

How exactly a solar storm would induce these massive changes on Earth is hard to discern, but one scientist tries in the following way:

> Plasma hitting the surface of Earth could heat and fuse rock, incinerate flammable materials, melt ice caps, vaporize shallow bodies of water creating an extended deluge of rain, and send the climate into a warming spell. The release of pressure that follows the melting of thousands-of-meters-thick ice sheets can induce earthquakes and even cause hot rock under pressure to melt and erupt to the surface as volcanoes. The world was in chaos. . .[40]

All well and good, but what about explaining the Younger Dryas which was a period of rapid *cooling* that followed this alleged solar mass ejection?

[38] Kasha Patel, "Massive prehistoric solar storm," Washington Post, 2023, Oct 11, https://www.washingtonpost.com/climate-environment/2023/10/11/largest-solar-storm-research-earth/

[39] Alex Wilkins, "Largest known sola storm struck 14,300 years ago." New Scientist, 9 Oct 2023. https://www.newscientist.com/article/2396456-largest-known-solar-storm-struck-earth-14300-years-ago/

[40] Robert Schoch, "Solar outbursts and the end of the last ice age," Research Highlights, https://robertschoch.com/plasma_iceage.html

Robert Schoch, a professor from Boston University, also believes a huge solar storm impacted Earth. He speculates about the dynamics. In a personal email I received from him, he puts things in a remarkable perspective. He says that a cosmic event at the beginning of the Younger Dryas (14,000 years ago) caused "a massive glacial flood, caused by solar activity melting ice dams and glaciers, changed ocean circulation patterns in the Atlantic, thus initiating the Younger Dryas cooling. The end of the Younger Dryas resulted from a solar outburst causing dramatic warming."[41] [42]

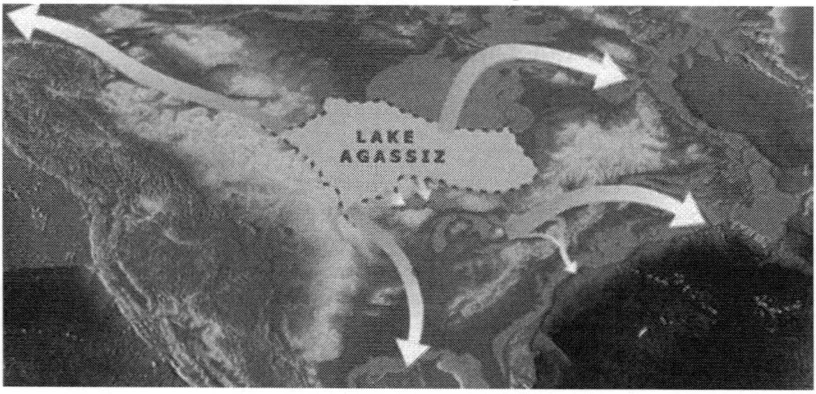

Lake Agassiz

In glacial times, there was a giant lake covering what are now the Great Lakes called Lake Agassiz. When these cataclysms happened, the ice dam burst and the lake flooded virtually most of North America. Here is a map of that disaster, followed by a description of its effects:

> One of the most significant events associated with Lake Agassiz was the catastrophic flooding that occurred when its ice dam broke. The ice dam, which formed at the southern end of the lake, was created by the accumulation of glacial ice that prevented the lake from flowing into the Mississippi River. As the water level in Lake Agassiz rose, the pressure on the ice dam increased. Eventually, the dam gave way, releasing a massive flood that swept across the northern United States and Canada. The floodwaters carved deep channels in the landscape, creating features such as the Grand Coulee and the Columbia River Gorge. The effects of the Lake Agassiz flood were felt as far south as

[41] personal email.

[42] See also Sam Nobel Museum, "Pleistocene Extinctions," Oklahoma Museum of Natural History, No date, https://samnoblemuseum.ou.edu/understanding-extinction/extinctions-in-the-recent-past-and-the-present-day/pleistocene-extinctions/

> the Gulf of Mexico, where sediment deposits from the flood were found. The floodwaters also altered the course of several major rivers, including the Missouri and the Mississippi.[43]

This is a very good time for the reader to watch Graham Hancock's episode of *Ancient Apocalypse* where he very dramatically illustrates how gigantic these floods across North America were.

This analysis helps us all understand more about the dynamics of these cataclysms, but it seems we might be talking of two separate chaotic events if we start trying to put all this academic rhetoric into some consistent frame.

Let us examine that concept next.

[43] Joseph Shavit, "The unrivaled majesty of North America's sixth Great Lake," The Brighter Side of News, 15 April 2020https://www.thebrighterside.news/post/the-unrivaled-majesty-of-north-america-s-sixth-great-lake

Could it have been two catastrophic events, not one?

From the graph of temperature, below, it would appear that something suddenly happened about 14,000 years ago. The temperature soared and then fell back. Then it reared up again with another gigantic temperature increase in the Younger Dryas.

Two events. In this figure the events are labeled strike 1 and strike 2.

The sudden warming might have been due to a solar cataclysm that would have set off fires and caused warming, ice melt, floods, and then so much soot that the sun became occluded and cooling began. Thus begins the Bolling-Allerod Interstadial and a plunge into a cold period. Then a second event happened at 12,000 ybp, where the Younger Dryas is dated, and you can see the temperature spike upwards again.

We may be talking about two cataclysmic events, a massive solar event followed by either a comet or supernova hitting a thousand years later. The graph tries to sort this out more clearly.

The cave lion disappeared at almost the same time as the Woolly Rhino. [44]

Traditional arguments are that the 35 principal species that we know died off between 14,000 and 9000 ybp (years before present) because of either climate change or human hunter overkills.

But there is now an alternative hypothesis: Two cosmic cataclysms happened, a massive solar eruption followed either by a comet or supernova hitting the planet: two cosmic events, followed by two sets of mass extinctions.

[44] Anthony Stuart and Adrion Lister, Extinction chronology of the cave lion. Quarternary Science Reviews, Vol 30, August 2011, pp. 2329-2340.

https://www.sciencedirect.com/science/article/abs/pii/S0277379110001320

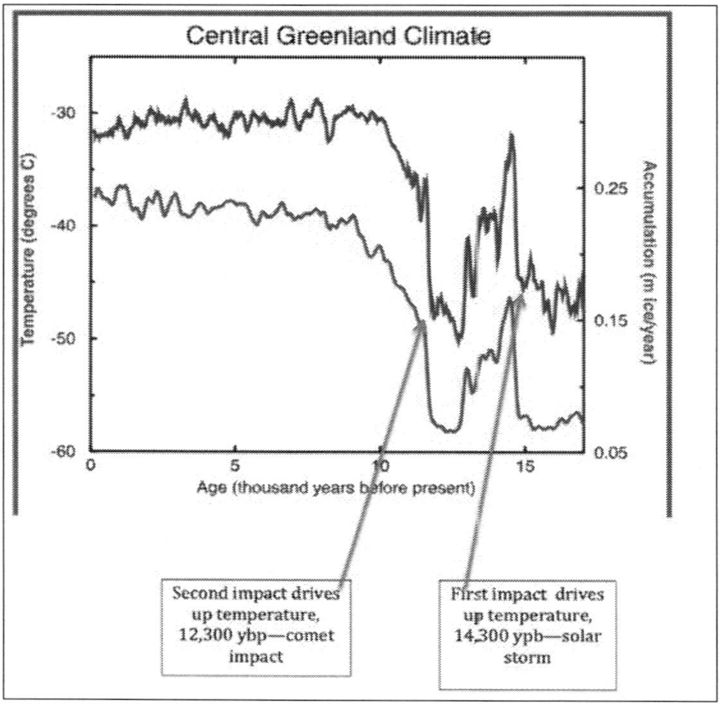
Central Greenland Climate
Temperature (degrees C)
-30
-40
-50
-60
Accumulation (m ice/year)
0.25
0.15
0.05
0
5
10
15
Age (thousand years before present)
Second impact drives up temperature, 12,300 ybp—comet impact
First impact drives up temperature, 14,300 ypb—solar storm

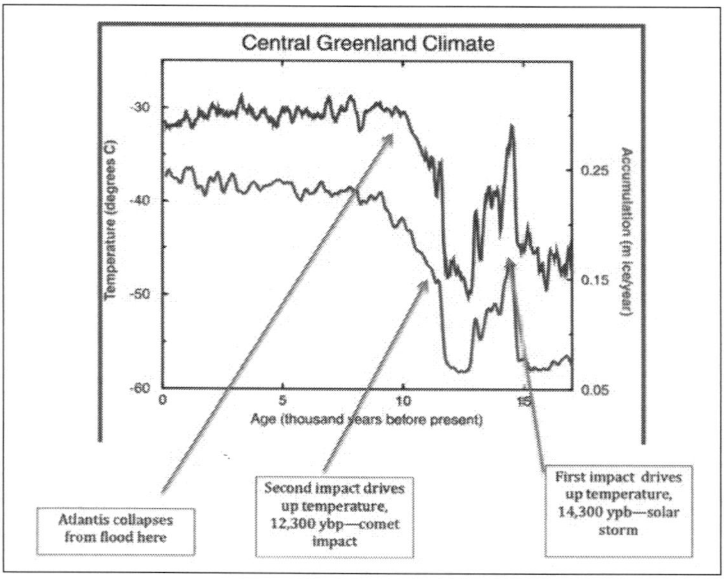
Central Greenland Climate
Temperature (degrees C)
-30
-40
-50
-60
Accumulation (m ice/year)
0.25
0.15
0.05
0
5
10
15
Age (thousand years before present)
Atlantis collapses from flood here
Second impact drives up temperature, 12,300 ybp—comet impact
First impact drives up temperature, 14,300 ypb—solar storm

The species that went extinct had survived many ice age climate fluctuations. The Woolly Rhino, for example, was strutting across this planet for 5 million years.[45]

Woolly Rhino, and cave lion went extinct 14,000 years ago, prior to the advent of the Younger Dryas

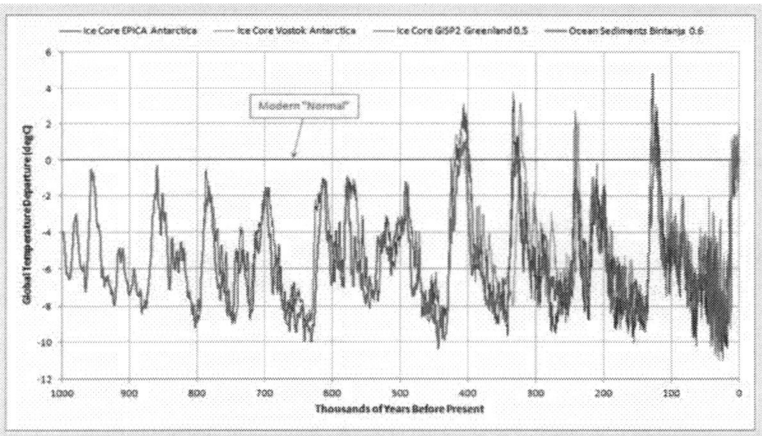

Global temperature reconstructions for the last million years from oxygen isotope ratio analyses of ocean sediment cores (Bintanja), Antarctic ice cores (EPICA and Vostok), and a Greenland ice core (GISP2). Click on graph to enlarge.

This graph shows global temperatures over the last million years. *At least four times*, global temperatures reached present modern-day levels (red line). In that time the Woolly Rhino, Woolly Mammoth, Cave Lion, and Saber Toothed tiger survived them all.[46]

[45] Britannica, "Wooly rhinoceros extinct mammal," Jun 5 2024https://www.britannica.com/animal/woolly-rhinoceros

[46] "Three million years of climate change," Climate Concerns 7 Dec, 2014, https://oz4caster.wordpress.com/2014/12/07/three-million-years-of-climate-change/

The cave lion roamed about for 370,000 years. If you plug that into this graph, he survived *at least 3 massive climate changes*. Why suddenly did he perish 14,000 years ago? [47] It wasn't climate change. *It was something else.*

Again, there seem to be two catastrophic events, a massive Miyake solar storm 14,300 years ago, followed two millennia later by either a comet or supernova. They rather perfectly define the spikes in temperature that mark the beginning and end of the Younger Dryas.

These might be the cataclysms that caused the extinctions. Certainly human hunter-gatherers can claim some "credit," but it is hard to stretch that hypothesis to account for all of it.

To repeat from earlier, if hunters killed off the large animals, and they are solely responsible, it only took them 3,000 years in the Americas to wipe out 80% of the megafauna, but 40,000 years in Europe and 40,000 years in Australia. Yet in Africa they had 190,000 years to kill them all, but they only managed to wipe out 21% of the species.

The mainstream argument makes no sense, but these data are quite recent discoveries and the significance has not yet sunk in!

Below you see a graph of the discovery, the investigators, all of whom have been cited, and the recent nature of their findings.

While traditional archeologists may resist hearing that their work is being challenged, the fact is that the leading hypothesis—that human hunters were responsible for the extinctions or that climate change brought about the die offs—is flawed to the extreme.

Still, the comet, supernova, and solar cataclysm theories are quite new and much remains to be learned. What helps make these explanations credible, however, is that they can explain spikes in radiocarbon levels, beryllium levels, the discovery of nanodiamonds, "microspherules" (siliceous scoria-like objects), platinum deposits, corundum, mullite and other rare meteoritic minearananols and melt-glass that only forms under high temperatures.[48] The hunter-overkill/climate change explanations fail entirely to explain any of that data. Proponents of the conventional theory totally ignore this data, and can't explain it; they don't even seem to want to.

But the comet–supernova–solar cataclysm hypotheses camp bases their arguments precisely on these data.

Did humans die too?

[47] New World Encyclopedia, Cave Lion, https://www.newworldencyclopedia.org/entry/Cave_Lion

[48] T. Bunch, et. al, "Very high temperature impact melt products as evidence for cosmic airbursts 12,900 years ago," PNAS June 18, 2011 https://www.pnas.org/doi/full/10.1073/pnas.1204453109

Table 2. Cataclysm discoveries and Investigators

Discovery	Investigators	Date of discovery or publication
Comet impact	Richard Firestone, Andrew Moore et. al. James Kennet, et. al.	2001–2007 2024
Supernova	Robert Brakenridge	2023
Solar storm (CME)	Eduoard Bard, Tim Heaton, et. al	2023

The obvious question we have yet to ask is, 'did humans die off like all the big animals during the Younger Dryas?' *If they didn't*, then maybe the overkill hypothesis has some truth to it.

However tantalizing to the mainstream, that idea has many problems: One study argues that the human population actually suffered a decline too.

> The results of each analysis suggest a significant decline and/or reorganization in human population during the early centuries of the YD, (Younger Dryas) varying in extent by region. Archaeological settings formerly heavily utilized, such as stone quarries in the south-eastern U.S. appear to have been largely abandoned, while over large areas, a substantial decline occurred in the numbers of diagnostic projectile points and cultural radiocarbon dates.[49]

One such group in North America was the Clovis people.

> About 13,200 years ago, near the end of the last Ice Age, the Clovis people arrived in North America. These Paleo-Americans were best known for distinctive stone points — called Clovis points — along with bone and ivory tools. Genetic testing of a known Clovis burial found that modern Native American populations seem to be closely related to the Clovis people. But the Clovis people seem to have abruptly vanished from the scene about 12,800 years ago.[50]

[49] D. Anderson, et. al., "Multiple lines of evidence for possible human population decline/ settlement and reorganization during the early Young Dryas, "Quaternary International, 14 October 2011, pp. 570-583. https://www.sciencedirect.com/science/article/abs/pii/S1040618211002333

[50] Eric Ralis, "Meteor may have led to the decline of the Clovis People," Earth.com, Sep. 2024 https://www.earth.com/news/meteor-decline-clovis-people/

More on the Clovis people

> The discovery of the Clovis culture in New Mexico pushed the settlement of North America back to what archeologists believed were the first people to cross a land bridge between Siberia and Alaska at the end of the last Ice Age. The ancient people then followed an ice-free corridor south and dispersed across the continent. For decades after the Clovis discovery, researchers thought that the Clovis people were the first Americans. Clovis people hunted a range of animals including deer, rabbits, coyotes, and birds. The ancient Clovis people were food opportunists that depended on gathering the edible seeds, nuts, and roots of plants, fishing, and hunting small mammals for sustenance as well as taking down some of the larger animals.[51]

Stone tools of Clovis people, Source: Wikipedia

While the Clovis people disappeared at the time of the Younger Dryas, their DNA survived and a recent study delineated their ancestry more fully. They seem to have come and gone and then been replaced by other gene pools:

> The large-scale population replacement is a process that was not widely expected by archaeologists," said David Reich, a geneticist at Harvard Medical School and the Howard Hughes Medical Institute who co-authored the study, in the *Cell* press release.[52]

Another study adds to our understanding:

[51] New Mexico True, "The Story of the Clovis People," 18, Mar, 2019, https://www.newmexico.org/blog/post/the-story-of-the-clovis-people/

[52] Becky Little, "Why did the Clovis People vanish?", History, 9 Nov, 2018 https://www.history.com/news/clovis-migration-discovery

> Michael Waters, distinguished professor of anthropology .... along with Texas A&M anthropologist David Carlson and Thomas Stafford of Stafford Research in Colorado, have had their new work published in the current issue of *Science Advances*. The team used the radiocarbon method to date bone, charcoal and carbonized plant remains from 10 known Clovis sites in South Dakota, Colorado, Pennsylvania, Ohio, Virginia, Montana and two sites in Oklahoma and Wyoming. An analysis of the dates showed that people made and used the iconic Clovis spear-point and other distinctive tools for only 300 years. "We still do not know how or why Clovis technology emerged and why it disappeared so quickly," Waters said.[53]

One suggestion, of course, is that a comet, supernova, or solar flare caused fires, floods, and with the mass extinctions of megafauna, so too did the Clovis people succumb to the same set of environmental circumstances. (By the way, the extinctions were not planet-wide. They were severe in North and South America, but Africa and South Asia experienced much lower rates.)[54] [55]

But other human groups seem to have perished or been seriously reduced during the Younger Dryas too. Author Jonathan Adams of the Oakridge National Laboratory says that in Europe a similar process unfolded: "The climate transition into the Younger Dryas seems likely to have eliminated much of the previous late Paleolithic population of northern and central Europe." [56]

A study in the Benelux area of Europe shows an abrupt population decline during the Younger Dryas. [57] A 2019 study also reports a severe population decline in South America, but that data appears to happen after the end of the Younger Dryas, not during it. [58]

[53] Keith Randall, "New Clues revealed about Clovis People," Texas A&M Today, 23 Oct. 2020, https://today.tamu.edu/2020/10/23/texas-am-expert-new-clues-revealed-about-clovis-people/

[54] Wikipedia. Late Pleistocene Extinctions, https://en.wikipedia.org/wiki/Late_Pleistocene_extinctions

[55] Photo which follows of 12 year old boy genome is credited to the following source: https://www.sci.news/genetics/science-genome-clovis-boy-01760.html

[56] Jonathan Adams , "The Younger Dryas and the distribution of Nostratic languages," Oakridge National laboratory, No date, Nostrathttps://www.esd.ornl.gov/projects/qen/Indo.html

[57] P.M. Vermeersch, "The human occupation of the Benelux during the Younger Dryas," Quaternary International, 15 Oct 2011 pp. 267-276https://www.sciencedirect.com/science/article/abs/pii/S1040618210004180

[58] Science News, "Abrupt Climate change drove early South American population decline," 9 May, 2011, https://www.sciencedaily.com/releases/2019/05/190509080039.htm

Note though that mainstream anthropology takes issue with all of this, the impact hypothesis, or any other cataclysm. It does not see a decline in the human population and is wedded to the hypothesis that the extinctions were caused by human hunter overkill, and that's that! So say 13 anthropologists in one single refutation article. [59]

But if you leave aside the internecine social sciences and the biases that Graham Hancock rightly accuses them of, you can find a group of geneticists and cellular biologists who studied chromosomal sequences in the Americas and said *there was a substantial loss of 'lineages' during this time and a "profound genomic impact."* This time there were 15 authors contributing, none social scientists to my knowledge.[60]

I decided to try to confirm if my interpretation of their article was accurate, so I wrote to one of the authors for a clarification. The reply went as follows: "I'm glad you are interested in my article. Your interpretation is correct. From Y chromosome genomic data, we estimate that during the Younger Dryas there was a decline in the human population..."[61]

So, yes, human population declines accompanied all the megafauna extinctions, and they happened together despite the carping of traditional archeology.

Objections overruled!

We will leave the issue of bias in anthropology, archeology, and the social sciences to a later portion of this text. For the time being let us conclude that by the close of the Younger Dryas, a gigantic number of species went extinct that had been here for hundreds of thousands of years, and that the human population suffered declines as well.

In the meantime, there is still more evidence that something deeply catastrophic happened on this planet beginning in 12,000 BC, and we need to look at this evidence too.

[59] Vance Holliday, et. al., "Comprehensive refutation of the Younger Dryas Impact Hypothesis," Earth Science Reviews, December 2023,https://doi.org/10.1016/j.earscirev.2023.104502

[60] Paula Paz, et. al., "Human Y chromosome sequences from Q Haplogroup reveal a South American settlement pre-18,000 years ago and a profound genomic impact during the Younger Dryas," PLOS One 17 Aug. 2022, https://journals.plos.org/plosone/article?id=10.1371/journal.pone.0271971

[61] personal email

Other evidence of massive cataclysms: bone-beds

Other curious datum suggestive of a massive cataclysm are bone-beds or repositories of vast quantities of separated and assorted bones from myriad animals all piled up in an area and fully separated from the skeletons from which they were attached. These are enormous numbers of bone fragments strewn all over as if the animals were torn to shreds and washed by a giant flood to finally wash down and settle in one funerary depository.

One such bone-bed is in Melbourne Florida. It contained bones of over 36 species of mammals plus birds, reptiles and even humans. The radiocarbon dates given were between 20,000 to 10,000 years, so it fits rather nicely into the Younger Dryas period.[62]

The bones of myriad species were scattered: camels, dire wolves, Florida cave bears, giant armadillos, giant beavers, giant bison, giant ground sloths, mammoths, mastodons, saber-toothed cats, and tapirs.[63]

One might want to ask how mammoths, camels, wolves, bears and beavers all ended up in the same spot, and certainly one answer is that some sort of catastrophic flood killed them and flushed their bodies downstream until they came to rest in this spot all mashed up together.

[62] Melbourne, University of Florida Vertebrate Fossil Locality BR002, https://www.floridamuseum.ufl.edu/florida-vertebrate-fossils/sites/melbourne/

[63] Melbourne bone bed, Wikipedia, https://en.wikipedia.org/wiki/Melbourne_Bone_Bed

Texas and Nebraska

One bone-bed is The Bonfire Shelter in Texas, Bone-bed 2. It contains remains of horse, camel, and mammoth bones dated to 12,460 years ago, right in the time period we have been wondering about.

Agate bone-bed in Nebraska (above), Bone-bed 2 in Texas (below).

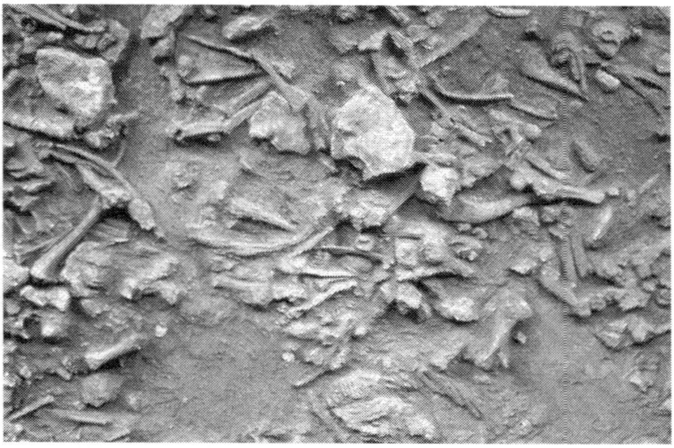

As you can see from the photo, these bone-beds contain twisted, broken fragments of bones of myriad animals, as if they had been tossed, broken apart and destroyed by some kind of massive event. Intact skeletons are not the order of the day.

Another bone-bed in Nebraska, is called the Agate Fossil beds, and you can again see myriad broken, dis-attached bones scattered everywhere.[64] [65] One of the Agate beds was dated at 10,000 years ago.[66] **The bone-beds in Nebraska were as much as two feet thick.**

An impressive bone-bed was found in Gonstsy, Ukraine, and the bones were similarly scattered about, but curiously the radiocarbon dates are almost exactly at the time of the first cataclysm which we have dated at 14,300 years. The dates for these scattered bones are 14,300-13,200 ybp. [67]

The Shelton bone-bed found in Michigan had bones dating to 12,320 years ago.[68]Also, single bone discoveries corroborate that something quite catastrophic was happening between 14,000 to 12,000 years ago: Woolly Mammoth fossils were found in England and dated to 14,000. [69] The Mount Holly mammoth of Vermont dated at almost precisely the time of the comet 12,882.[70]

I will not recite more individual extinctions, but it certainly corroborates that mass extinctions were unraveling during the Younger Dryas which started with a celestial cataclysm and ended with one too.

Conclusion

This concludes part I of this text. If we were to summarize the 'best evidence' and fashion a conclusion, it would be this: Two celestial catastrophes occurred which human beings and scholars for the last few centuries were unaware. Isaac Newton in the 1600s "suspected" there was a huge celestial event—he thought a comet—which caused massive flooding and

[64] National Park service, Agate Bonebed, https://www.nps.gov/agfo/learn/nature/agate-bonebed.htm

[65]The Agate bone-bed was dated at 10,300 ybp: if you look at the graph of temperature this is the very last moment when cold temperatures ended in a warming period that last to this day; Craig Lee, "Refining the Chronology of the Agate Basin Complex, Radiocarbon dating," Plains Anthropologist, August 2011 "https://www.jstor.org/stable/23209312

[66]Robert Neuman, "Radiocarbon dated archeological remains on the Northern and Central Plains," American Antiquity, Oct., 1967.https://www.jstor.org/stable/2694075

[67]The Gontsy Mammoth bone hut site " Don's Maps, 2014. https://www.donsmaps.com/gontsy.html

[68] J. Shoshani, et. al., The Shelton Mastodon Site, September 2005, Research Gate. https://www.researchgate.net/publication/30849295_The_Shelton_Mastodon_Site_Multidisciplinary_Study_of_a_Late_Pleistocene_Twocreekan_Locality_in_Southeastern_Michigan

[69]New radiocarbon dating information changes, , Dinosaur and Prehistoric Animal News Stories, 20, June 2009https://blog.everythingdinosaur.com/blog/_archives/2009/06/20

[70]Sharon Huntley, Wooly Mammoth Research form Mount Holly, 19 March 2021, The Vermont Journalhttps://vermontjournal.com/featured-articles/wooly-mammoth-research-from-mount-holly-dates-early-bones/

chaos on Earth, but he merely suspected it; he had no way to gather evidence to prove it.

Isaac Newton

It was only in the last decade that we started to finally find the clues we needed. The most likely theories, that is the theories that seem to have the best evidence, point to two cataclysms. The first was a massive Miyake sized solar storm that shocked the Earth 14,300 years ago. How can we be so precise? Count the fossilized tree rings!

That, in turn, set off fires—almost nine percent of all the vegetation burned to the ground—then floods, melted ice sheets, and a sudden massive increase in the temperature of the Earth followed. Greenland went from minus 58 degrees C to minus 30 degrees rapidly.

That was a huge shift. Fires or soot polluted the atmosphere and there was a collapse of the warming as the Earth plunged back into cold and darkness for almost two millennia. Then 12,900 years ago another celestial strike occurred, this time a comet, and it and/or its fragments once again started a sudden warming trend with a massive impact fusing rocks, starting fires and floods and warming the Earth rapidly to a level which we still have to the present day. It helped put an end to the ice age.

In the interval of time, extinctions of 35 different species occurred. They had been on Earth for hundreds of thousands of years and survived multiple ice ages, but these two events drove them into nonexistence: the Woolly Mammoth, mastodons, saber-toothed tigers, and a taxonomy of large animals. This extinction process continued well into 10,000 years ago.

Humans also perished during this time, and the Clovis people of North America vanished.

Archeology and anthropology seem opposed to these findings.

But these two professions increasingly seem sclerotic, insisting that there was no comet impact, no solar cataclysm, obdurately proclaiming all extinctions were caused by human hunter-gatherers as they spread across the planet.

The species extinctions congregated in the time band between these two cosmic catastrophes, and the human population suffered as well during this period. So that completes our first sojourn into the undiscovered past of our species and the planet. With that behind us, there is much more to address.

PART TWO. HUMANITY THROUGHOUT THE YOUNGER DRYAS

Myth, Memory, and Mnemonics

As the sojourn into the planetary past is behind us, I'd like to remind you that I am a psychologist and not competent to do anything more than report what I read in the scientific journals of astronomy, anthropology, solar physics, and cell biology. That stated, there is a considerable bit of psychology to examine in all this, and, frankly it resonates most with Joseph Campbell and Carl Jung.

I promise to get back to these extinction events, but let us add in a bit more of the human dimension.

Joseph Campbell taught mythology and ancient classic literature at Sarah Lawrence College and later and became a darling of the liberal intelligentsia before his death. He was strongly influenced by Jung, and his contribution fundamentally was that mythology was not a series of irrational made-up stories with no roots in reality but that it pointed to something, often something very real.

Take Noah's Ark as an example. The biblical story was written in about 1500 BC, and despite searches looking for his ark, the mythic nature of the story seems lost. The story's meta-message is that there was a great flood and many animals faced death and extinction. Remind you of anything?

Well, there was no worldwide flood in 1500 BC that we can reliably fix, [71] but there was a worldwide flood about 8000 years before, where waters in the ocean rose close to 400 feet. Graham Hancock says that the melting ice age submerged 10 million square miles of the Earth.

> Meanwhile, as the Earth warmed, the Mediterranean Sea, reduced to a huge hypersaline lake when low global sea levels cut it off from the Atlantic during the Younger Dryas, slowly refilled and around 8,200 years ago swamped over a silt dam that had plugged the Bosphorus Strait between Europe and Asia. The overflow became a raging torrent that, within a few days, cut a deep channel through silt plugging the strait. With the weight of the newly filled Mediterranean behind it, the water cascaded 150 metres into the Black Sea.
>
> The massive waterfall, estimated at 300 times the volume of Niagara Falls, would have created a deafening roar that could have been heard 500 kilometres away. Submarine imaging surveys have revealed the geometric outlines of ancient settlements on the margins of the former lake shore that drowned as the lake's rapidly rising waters advanced several kilometres a day across its gently sloping shores. Historians suspect the Black Sea flood 8,200 years ago was the source of the biblical tale of Noah's Ark.[72]

Another source suggested that a great tsunami hit the Mediterranean and the coast of Israel about 10,000 years ago:

> The large early Holocene tsunami occurred between 9.91 to 9.29 ka with an extreme ~16 m wave height and 1.5–3.5 km run-up on the coast of NW Israel. Early Pre-Pottery Neolithic sites were destroyed along the Carmel coast creating a ~4,000-year settlement gap in the Neolithic archaeological record for the area.[73]

But Noah's Ark, as a mythic story, has to be understood differently. It's not that we are looking for the remnant of his boat, but that we are looking for the meta-meaning of the story and whether that has any roots in reality. As I have written previously,

[71] "Yes, Noah's flood may have happened, but not over the whole Earth," Reports of the National Center for Science Education, Sept 2009, https://ncse.ngo/yes-noahs-flood-may-have-happened-not-over-whole-earth

[72] Farmageddon Ibid., https://www.theage.com.au/technology/farmageddon-20070819-ge5mar.html

[73] Ibid.

> The flood myth is not limited to the Bible. It is a legend that is found in Aztec mythology, in Central American tradition among the Mechoacanesecs, in the epic of Gilgamesh fifteen centuries before *Noah's Ark* was transcribed in the Old Testament, and in Indian tribes of Ecuador.

As the flood myth repeats so too does the figure of Noah.

> More than 500 deluge legends are known around the world, and, in a survey of 86 of these (20 Asiatic, 3 European, 7 African, 46 American and 10 from Australia and the Pacific. . .researcher Dr. Richard Andree concluded that 62 *were entirely independent* of the Mesopotamian and Hebrew accounts.[74]

> If there are sixty-two independently arising tales, all of which have a very similar deep structure, the myth could be interpreted *as referring to a real event.* In fact, it has been suggested that the ubiquity of this story may point to the retreat of the massive ice sheets that covered the planet. From 11,000 BC until 9,000 BC a mile-thick ice cap receded. Ice age mammals were found as far south as Texas. The graphic of the temperature of the Greenland ice sheet shows a rapid rise in temperature about 11,000 BC. Thousands of animal species became extinct, and the oceans rose 350 feet. Deglaciation-induced flooding may be a realistic explanation for all the consequent flood myths. [75] So, is it not possible that the universality of the flood myth can be understood as a human archaic memory of a natural threat to man's survival?[76]

More to the point, in 2024 an article appeared in the journal *Archeology* describing an almost exact replica of the Noah deluge story, but it occurred in Mesopotamia, a millennium earlier. Here is the story from the Gilgamesh tablet, written a thousand years before the Noah narrative:

> The tablet describes how the gods sent a flood down to destroy Earth. However one god, Ea, alerts Utu-napishtim, the ruler of an ancient kingdom, of the plan and instructs him to build a boat to save himself and his family along with "birds and beasts of all kinds," according to the British Museum, which counts the artifact as part of its permanent collection. Just like in the Book of Genesis, the voyagers release birds to see if the waters had receded enough for land to emerge.

[74] Hancock, Ibid., p. 193.

[75] Source for the graph on temperature over Greenland R.B. Alley "The Two-Mile Time Machine," Princeton University Press, Oct. 2014

https://press.princeton.edu/books/paperback/9780691160832/the-two-mile-time-machine

[76] J. Kroth, *Aliens and Man*, New York: Algora Publishers, 2001, p. 7

Later, Utu-napishtim tells Gilgamesh of his experience, according to the text.[77]

So the story of Noah's Ark is mythic, yes, but one *that preserves a memory trace* dating to thousands of years earlier. It can be construed as *a very true story* about a massive flood that engulfed the whole of the Earth. And there are 62 other mythic versions communicating the same message forward. They are narratives pointing not to mythical make-believe, but a fundamental reality: There was a great cataclysm that threatened all life on Earth, including human lives. In fact, 35 species went extinct in this unraveling cataclysm—despite Noah's attempt to save as many as he could!

So, yes, psychology does play a role in this discussion.

Myths which lead to facts

Another example of this idea is the story of Arthur Evans, a British archeologist.

"If we hold to the politically correct view that myths are delusions with no referent in reality, we should pause to remember the story of Troy described in the Iliad. The accounting was also deemed a fantasy until Heinrich Schliemann believed Troy to be a real place. With a little digging he proved the myth was an accurate record and a bona fide collective memory.

Sir Arthur Evans, British archeologist

Sir Arthur Evans, fascinated by Theseus killing the Minotaur, and disbelieving his contemporaries who said such stories were mere loquacious fantasy, unearthed Knossus and the Minoan civilization of Crete. This story of discovery repeats many times over as archeologists uncovered Nineveh, the city of Ur, Babylon, Sumer, Acadia, the ancient birthplace of Abraham, the kingdom of King Sargon II, even Babel... all thought to be the stuff of fables, legends, and allegories instead of concrete collective recollections of real places stored in the archives of mythology.

[77] Jeffinfer Naweliki, "Gilgamesh Flood Tablet," Live Science, 15 Jun, 2024, https://www.livescience.com/archaeology/gilgamesh-flood-tablet-a-2600-year-old-text-thats-eerily-similar-to-the-story-of-noahs-ark

'One hundred and fifty years ago, much of the history in the Old Testament was considered pure fiction, including the existence of Sumeria (the biblical Shinar), Akkad, and Assyria. But those forgotten pieces of our past were discovered in the late nineteenth and early twentieth centuries when Nineveh and Ur were found."[78]... " 'Myths' of the Bible that referred to a number of legendary ancient cities and peoples were slowly proved to be factual history.'[79] "

So myths sometime lead to discovery of real things, and sometimes myths are just myths, and a cigar is just a cigar. Let us, therefore, enter this new territory with some enthusiasm.

What about Atlantis?

Image of Atlantis

We might ask 'just how far are we going to go with this?" The story of Atlantis is still obviously a myth, and any serious investigation of it is considered nutty and 'fringe.'

But if we follow out the leads of Joseph Campbell and Arthur Evans, namely that myths convey a deeper meaning and sometimes lead us to concrete realities, we have a myth staring at us saying that a great civilization was destroyed by the very same floods that we suppose were happening at the close of the Younger Dryas.

Don't forget those bone-beds where all these dismembered animals were scattered, torn to shreds, and deposited throughout the world. And while

[78] Will Hart, Ancient Agriculture, in search of the missing links, in Forbidden History, Ibid, p. 195.

[79] Kroth, Ibid., p. 12 see also Will Hart, Ibid., p. 71.

we're at it, note that whale bones were found in Michigan and Ontario dated to about 11,000 years ago. "What the hell were whales were doing in Ontario," you might ask? If we assume a flood caused by an ice age melting and a solar Miyake event, it becomes easier to place whales in these spots too.[80]

So let us gather up some evidence about Atlantis and see if anything passes the smell test. Is this a myth like Noah's Ark which will lead us to some amazing reality, or is the cigar just a cigar?

First, Plato wrote of it about 360 BC in the *Timaeus*, and he had learned of it from others. The story is traced to Egypt. Let's do some dating. Plato said Atlantis perished in a great deluge 9,000 years before him, so that makes it 9.000 years before 360BC, or 9360 BC, in other words, 11,384 years ago. That puts it rather close to the Younger Dryas.

Let us plot that on our Younger Dryas chart and see what it looks like:

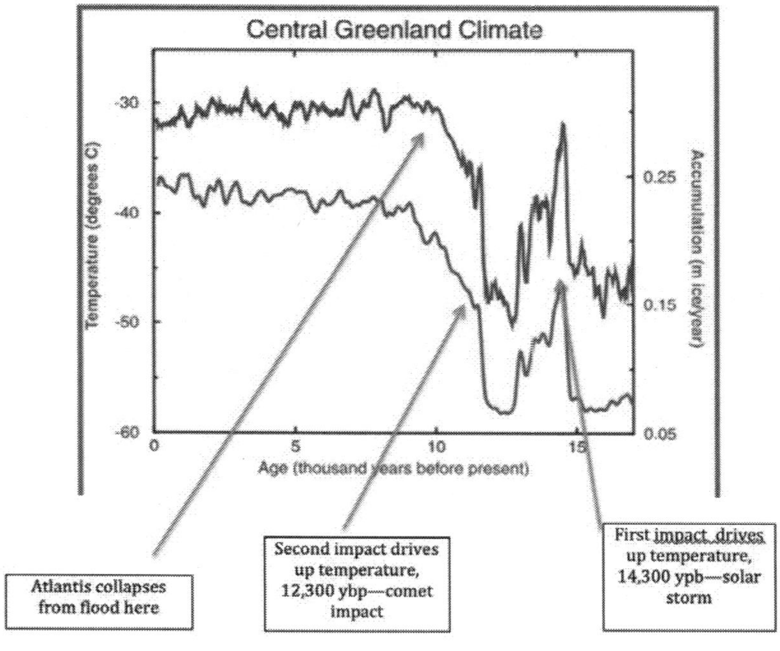

[80] Neil Carlton, Prehistoric Pakenham Whale on Display, the Milstone, November, 2017, https://millstonenews.com/prehistoric-pakenham-whale-on-display-in-almonte/

You can see that if indeed a cometary impact happened and drove up temperatures 12,300 years ago, creating a massive ice melt, floods, and enough chaos to raise the ocean 400 feet, Atlantis at 11,300 years certainly fits with the anomaly of the Younger Dryas and the ongoing extinctions.

Maybe just a coincidence, but clearly while these other species were experiencing a die-off, an advanced human culture might also have succumbed to the planetary chaos, just like the Clovis people.

Some factoids about Atlantis mythology

Ignatius Donnelly wrote a lengthy treatise a century ago entitled *Atlantis: the antediluvian world.* A Jesuit, his scholarship is voluminous. Here are a few items of interest that come from his life's work.[81]

- There is a reference in the mythology to the sun being responsible for "a great conflagrations upon the Earth." (p. 6) Could this be the Miyake event?
- The island of Atlantis was a "wonderful empire, which had rule over the whole island and several others as well as over parts of the continent."
- "There were violent floods and in a single day and night of rain all your warlike men in a body *sunk into the earth* [emphasis mine], and the island in like manner disappeared and was sunk beneath the sea." (p.8.)
- The whole country was described as being very lofty and precipitous on the side of the sea, but the country immediately about and surrounding the city was a level plain itself rounded by mountains which descended toward the sea; it was smooth and even but of an oblong shape extending in one direction three thousand stadia." (roughly 340 miles)
- The complexity of Atlantis was described as being beyond the capability of human ingenuity or craftsmanship, (p.12.)

[81] Ignatius Donnelly, Atlantis: the Antediluvian World Dec 2011, Amazon. https://www.amazon.com/Atlantis-Antediluvian-World-Ignatius-Donnelly/dp/0486233715

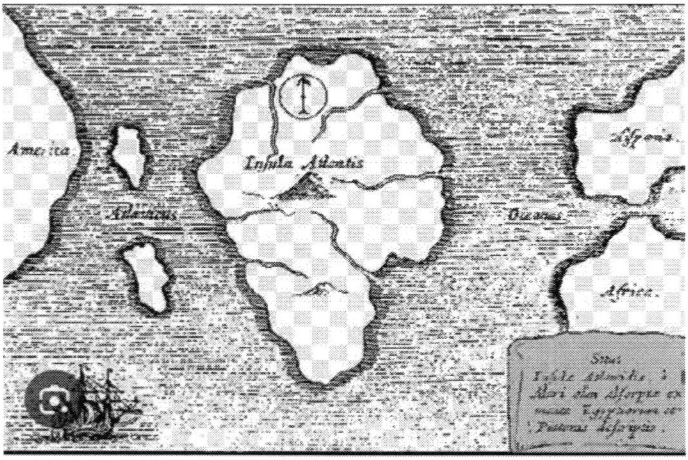

- "There are in Plato's narrative no marvels, no myths, no tales of gods, gorgons, hobgoblins or giants. It is a plain and reasonable history of a people who built temples ships, and canals; who lived by agriculture and commerce, who in pursuit of trade reached out to all the countries around them." (p. 15)

- The Atlanteans subjugated the whole of Europe and Africa as far as Egypt.

- They domesticated horses, ox sheep, goat, and developed cultivation of wheat oats, rye, and barely. (p. 16)

- Atlantis' reign extended for thousands of years.

- "The trees bore fruit at all seasons of the year, and it was located several stadi from the Pillars of Hercules" (a stadia is 185 meters)

- The destruction of Atlantis occurred in 9,600 BC (11,600 years ago)

Donnelly's book also provided a very unusual coincidence. He alleges Atlantis influenced all of Europe, Asia and much of South America. He spends a lot of time showing coincidences, especially in myth. Stories that are almost identical between Native American tribes and the Irish are examples. But he also holds that technology and stone-cutting techniques were disseminated from Atlantis; thus, he argues, the pyramids in Egypt and the muscular walls of Peru all come from the influences of Atlantis.

Right angled corner stones in Egypt and Peru

While much of this seems preposterous, there are artefacts we still see today that are otherwise difficult to explain. When I traveled to Egypt, I marveled at one right-angled corner stone. Strangely enough, when I visited Peru a few years later, I ran into the very same type of corner stone. Donnelly says Atlantis disseminated its technology both to Egypt and the Americas.

The location of Atlantis

Plato said that Atlantis was located beyond the 'pillars of Hercules' which is commonly known as the Strait of Gibraltar; in other words, somewhere beyond that, in the Atlantic Ocean.[82]

In 1669 Athanatius Kircher, a German mathematician, [83] drew a map of Atlantis, placing it right outside the strait of Gibraltar.

Since today there is nothing there, other than the Canary Islands, the Azores, and Portuguese Madeira, we have to scratch our heads in wonder. Just how fantastic is the story of Atlantis?

But what if we imagine ourselves raising 10 million square miles of submerged land—submerged due to the ice melt, i.e., the 'deluge'? What the Atlantic would have looked like 400 feet shallower? Actually, the U.S. Navy

[82]Wikipedia. Pillars of Hercules. https://en.wikipedia.org/wiki/Pillars_of_Hercules
[83]Wikipedia Anthanasius Kircher, https://en.wikipedia.org/wiki/Athanasius_Kircher

published a map of the Atlantic pre-deluge (antediluvian).[84] A clearer rendition can be seen below:[85]

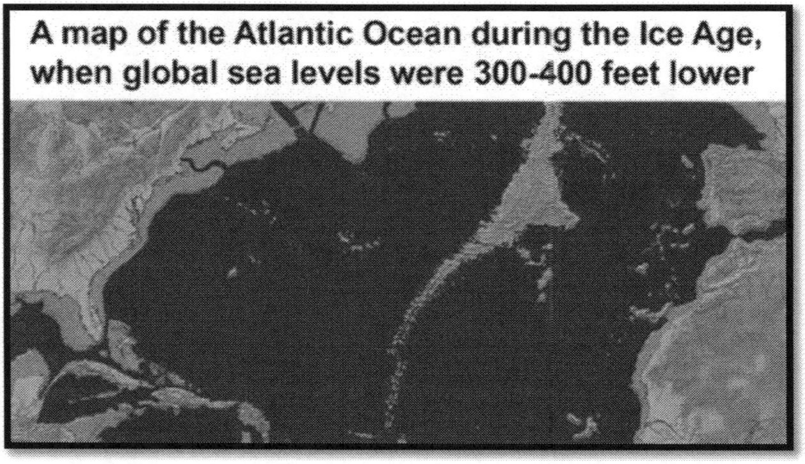

Buried somewhere beyond the Azores is where this hypothetical, still mythic culture may have existed. Actually, there is no compelling evidence that the Azores harbored any antediluvian advanced culture, but one geologist, Christian O'Brien, suggest the Azores was indeed much larger in those days.

> He concluded that the island had originally measured 720 km across from east to west, and 480 km from north to south, with high mountain ranges rising over 3660 metres above sea level. Before or during its submergence, it tilted by about 0.4° with the result that the south coast sank about 3355 metres but the north coast only some 1830 metres. Only the mountain peaks remained above the waters, and now form the ten islands of the Azores. O'Brien thinks the island could have sunk within a period of a few years or even months. . .[86]

[84] WikiUniversity, Continental Shelves, https://en.wikiversity.org/wiki/Continental_shelves/North_east_American

[85] A map of the ocean during the ice age when global sea levels with 300-400 feet lower, Mempedia, https://ifunny.co/picture/a-map-of-the-atlantic-ocean-during-the-ice-age-LE4JJHlCA

[86] Carolyn Silver, "The Mid Atlantic Range," Atlantic Online, 28 July 2008, https://atlantisonline.smfforfree2.com/index.php?topic=10182.75

One curious footnote: Plato estimated the size of Atlantis to be 555 km in width; O'Brien has it at 480. The erstwhile estimate is based on 2,300 year old mythology; the latter on geological assessments in this century. [87]

Another archeological study looked carefully at the Atlantic Ridge and seems to have found very interesting areas:

> It was now possible to visualise a great island about the size and shape of Spain, with high mountain ranges rising over 12,000 feet above sea level and impressive rivers running in curving valley systems. In the southeast, a feature which we have called 'The Great Plain' covered an area in excess of 3,500 square miles, and was watered by a river comparable in size to the River Thames in England. It has, and we shall sea, points in common with the great plain described by Plato in his Critias, as being a feature of the Island of Atlantis. [88]

An absence of archeological discoveries

Atlantis was large and powerful, said Plato. There were 10 kings governing 10 separate regions, and Atlantean power extended well into Europe and North Africa. But so far, there are no archeological artifacts to support these suppositions.

Most archeologists can't find anything resembling human habitation of the Azores until about AD 700, much too late for any Atlantean culture to have existed there. [89] Indeed, when the Portuguese landed there in the 1400s, there were no inhabitants. But we can keep looking:

Bimini Road

There is an underwater structure off the coast of Bimini, which is in the Bahamas off the coast of Florida. The so-called "road" is about half a mile long and composed of rectangular rocks which some call "pavement." It is in about 18 feet of water.

Many "Atlanticists" claim it is part of ancient Atlantis, and if you look at the Kircher map of 1669 (next page), there is a mysterious little island almost

[87] Christian & Barbara Joy O'Brien, The Shining Ones, Kemble, Cirencester: Dianthus Publishing, 2001, pp. 435-42; 'Survey of Atlantis', http://www.goldenageproject.org.uk/survey.html.

[88] Proof of the existence of a large island on the Mid-Atlantic Range, The Golden Age Project, 2024 https://www.goldenageproject.org.uk/342proof.php

[89] Bob Yirka, "Evidence of people on the Azores Archipelago 700 years earlier than thought,", Phys.org Oct 5, 2021. https://phys.org/news/2021-10-evidence-people-azores-archipelago-years.html

exactly where Bimini might be. I'll reproduce that drawing again for you. You can see Bimini off to the left.

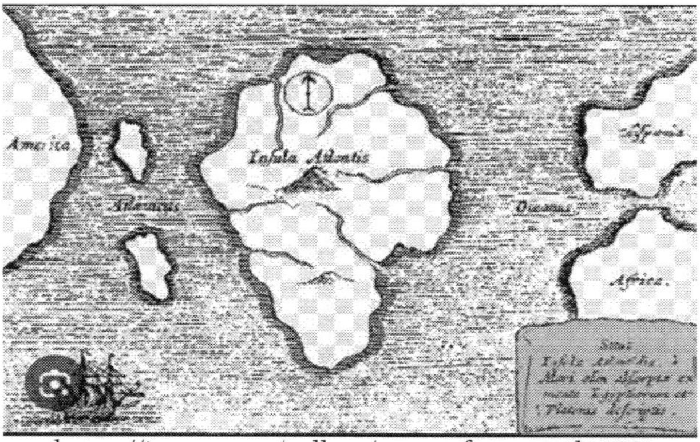

https://imgur.com/gallery/rows-of-stone-2l27RZm

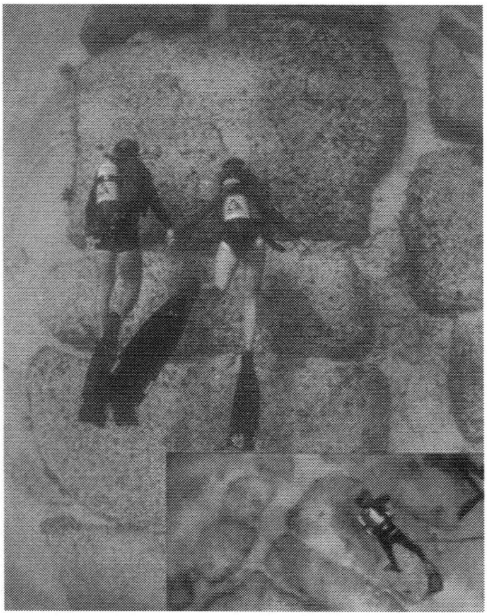

Underwater Bimini road structure

Source: https://imgur.com/gallery/rows-of-stone-2l27RZm

The Bimini road has been subject to archeological debate, almost all of which says either that it is not man-made or that it is much too early for Atlantis — only about 3,000 years old.

However, one study using a uranium-thorium method dated it at roughly 14,900 years which is rather exactly the time period we are looking at. But controversy surrounds this date, and this measurement tool, and many scholars take issue with it. [90]

In looking at the archeology of the Azores, I saw nothing that would support an advanced civilization's presence despite dozens of claims and YouTube videos to the contrary.[91]

What is interesting in this story is that if Atlantis existed, then human culture and civilization existed too *before the flood*, and just like the 35 species wiped out during the Younger Dryas, this society of human beings also succumbed.

A strange coincidence

I've never taken to Edgar Cayce, the "medium" who fell into a sequence of trances and then verbalized lengthy treatises on various subjects, one of which was on Atlantis. I always ignored that subject matter, but Cayce's book on Atlantis required I overcome my resistance. The book is, in my mind, a compendium of blather about astrology, God, purification, sex, and a Holy Spirit figure he called 'the entity'. [92]

He described Atlantis as a string of islands in the Atlantic whose influence and trade went from Honduras and Guatemala all the way to Europe and Africa. The only thing that caused a startle reaction in me was that he described it as involving Bimini. He said that in 1933, but the Bimini Road was not discovered by a diver until 1968, thirty-five years ahead of his time. I can't say anything more definitive from Cayce's work than report that uncanny coincidence. It is otherwise astrological babble.

Conclusion

About 57% of Americans believe the story of Atlantis is true. [93] But academic archeology in its stentorian ways believes all of that is nonsense,

[90]Wikipedia, Bimini Road. https://en.wikipedia.org/wiki/Bimini_Road

[91]Callum Hoare, Atlantis Exposed? Historians stunned by man-made structures breachign Atlantic, Jan. 2019, Express. https://www.express.co.uk/news/weird/1073103/atlantis-exposed-lost-city-historian-stunned-mountains-atlantic-spt

[92]Edgar Cayce, Atlantis, A.R.E. Press, 2009

[93]Miranda Leer, History of the Lost City of Atlantis History Defined, 2024, https://www.historydefined.net/history-of-atlantis/

pseudoarcheology, and as one put mainstream archeologist put it, "The scholarly consensus...is very clear: Atlantis was not a real place." [94]

Like it or not, we have been tip-toeing into pseudoarcheology and having the temerity to take Atlantis seriously, at least for a moment, and this will probably be subject to the same ridicule Arthur Evans felt.

The first problem, though, if we are honest with ourselves, is that we don't have any significant data which proves Atlantis ever existed, that is, which is beyond mythological accounts. And the Bimini Road is just a wild conjecture.

Until someone finds some stunning archeological edifices 400 feet underwater, we are driven to conclude either that Atlantis never existed, or we still have to keep looking. My apologies to academic archeology, but I find no harm in looking.

I do have one footnote to all this, and it is quite subjective, so I hope you will permit a personal observation: If you remember, the right-angled corner stones of Egypt and Peru are rather difficult to explain. Peru is a very long way from Egypt. Since there was no Panama Canal, one would have to sail around the horn of Africa and South America to reach Peru, a trip of at least 15,000 miles. It is hard to picture Egyptians spreading their technology to the pre-Inca peoples of Peru where these stones are hewn in the same way as the Egyptians as they traversed the waters of Cape Horn and the Cape of Good Hope to reach Peru.

But, instead, if we think along the lines of Ignatius Donnelly that Atlantis had colonies stretching from the Americas all the way to Egypt, then we can see that the Egyptians got their stone-cutting techniques from Atlanteans and so did the Peruvians. The distances for dissemination of this technology are more than cut in half:

Circumstantial evidence

Mark Carlotto in *Before Atlantis* makes a conjecture that that curious photo from NOAA of the bottom of the Atlantic looks like it could be a human construction. Unfortunately, it is a mile deep and not likely caused by a 400-foot rise in the ocean during the recession of the ice. Further, he thinks the photo may be an anomaly of the photographic process. Curiously, though, it is located pretty much where Atlantis was supposed to be, just outside the pillars of Hercules (Gibraltar).[95]

[94] Candida Moss, "Atlantis which no serious historian thought existed," The Daily Beast. Sept 2021, https://www.thedailybeast.com/atlantis-which-no-serious-historian-thinks-existed-is-making-people-insane-on-twitter

[95] Mark Carlotto, Before Atlantis, Amazon, Oct. 2013 Amazon,

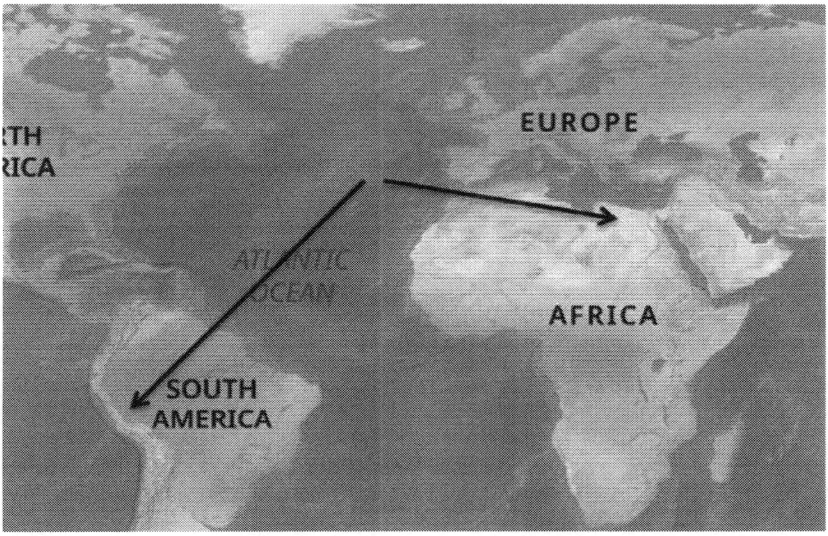

Distance from Atlantis to Peru and Egypt

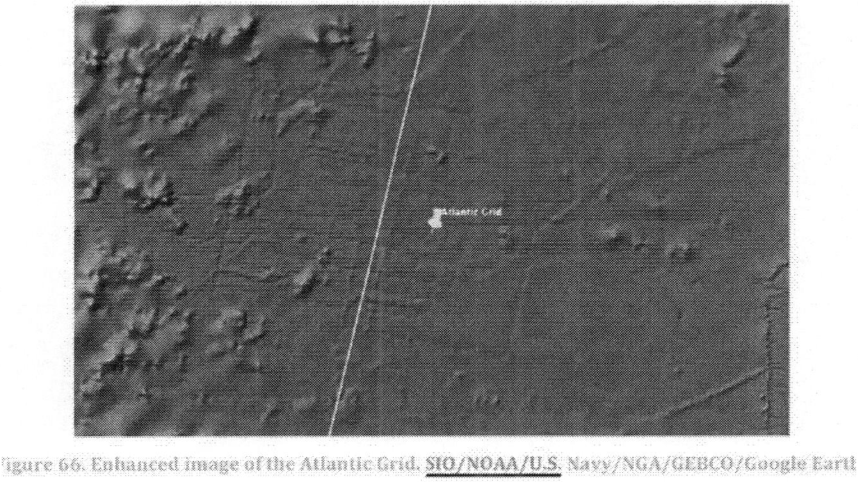

Enhanced image of the Atlantic Grid. SIO/NOAA/U.S.Navy/NGA/GEBCO/Google Earth

Photo shows a grid-like structure at the bottom of the Atlantic. Below its location just beyond the rock of Gibraltar

More circumstantial evidence: The lost city of Cuba

A recent discovery 650 meters under the water is 'the lost city of Cuba.' It was only discovered in the year 2000 and represents a series of smooth blocks which like look like granite, some up to 10 feet in size. They appear deliberately stacked on top of each other; others are circular.

The mystery is that these structures are too deep to be explained by rising sea levels. Since there are no volcanoes in Cuba researchers are scratching their heads for an explanation.

Other studies show that this formation covers an area of 2 square kilometers. Pauline Zalitzki, a marine engineer, said,

> It's a really wonderful structure which looks like it could have been a large urban centre. However, it would be totally irresponsible to say what it was before we have evidence.[96]

> The Mayans have a legend of a city that was swallowed up by the sea. Furthermore, some say it appears to be at least 6,000 years old.[97] Mayans emigrated there, and some were slaves, but there are no Mayan monuments in Cuba, so, if this is Mayan, it stands alone.[98]

Images of the 'lost city' under water.

[96] Cuban underwater formation, Wikipedia https://en.wikipedia.org/wiki/Cuban_underwater_formation

[97] "Team finds 6000- year-old city underwater in Cuba, Deseret News, Dec, 2001 https://www.deseret.com/2001/12/8/19620752/team-finds-6-000-year-old-city-under-water-near-cuba/

[98] Jason Yaremko, "Mayas and the Mesoamercian presence in Cuba," Oxford Academic, 16 July 2016 https://academic.oup.com/florida-scholarship-online/book/30357/chapter-abstract/257399379?redirectedFrom=fulltext

Location of 'lost city'

> The Mayan empire instead was located in Chiapas and Yucatán, now part of southern Mexico, through Guatemala, Honduras, Belize, and El Salvador and into Nicaragua.[98]

Not Cuba! These are "megaliths" of a kind you'd find at Stonehenge or Easter Island," said one of the team that discovered them.[99]

> "These are "megaliths" of a kind you'd find at Stonehenge or Easter Island," said one of the team that discovered them.[100]

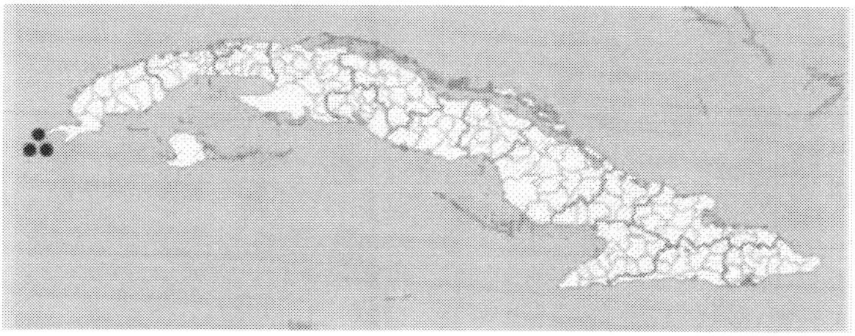

> The major problem for geologists and for archeologists is "how a city could have been built close to 700 metres (2,300 feet) below sea level today. . .even during the maximum drawdown of the world's oceans."[101]

A Florida archeologist says these structures are out of time and out of place.[102] Cuban archeologist, Manuel Iturralde, studied the area. I was able to contact him through email. He sent me this article:

> The megalithic stones are very unique structures.... I do not have any easy explanation for them in a natural geological process." Part of Dr. Iturralde's problem is that the stones appear to be cut granite blocks from 6.5 to 16 feet in length. Yet there is no granite in that region of the Caribbean, only limestone. A trip with a beefed-up ROV

99 "Team finds 6000-year-old 'city' in Cuba," Deseret News, Dec 2001, https://www.deseret.com/2001/12/8/19620752/team-finds-6-000-year-old-city-under-water-near-cuba/

100 "Team finds 6000-year-old 'city' in Cuba," Deseret News, Dec 2001, https://www.deseret.com/2001/12/8/19620752/team-finds-6-000-year-old-city-under-water-near-cuba/

101 Harriet Brewis, "Sunken lost city of Cuba turns rules of civilization on its heads," Indy 100, https://www.msn.com/en-ca/news/world/sunken-lost-city-of-cuba-turns-rules-of-civilization-and-geology-on-their-heads/ar-AA1nnQvR?ocid=msedgdhp&pc=U531&cvid=

102 Wikipedia Cuban Lost City, https://en.wikipedia.org/wiki/Cuban_underwater_formation

capable of drilling into the blocks to determine their composition was set for this July (2002). The National Geographical Society(NGS) had sent investigators, and then had pledged 2 million to the effort. But the needed funding has not materialized. The NGS continues to express interest, but there appears to be pressure from the State Dept. against NGS involvement with Cuba (partners with ADC).[103]

Dr. Iturralde said funding ran out, and more research is needed to determine much of anything. I remarked that since there are no Mayan ruins in Cuba, it would be strange if this underwater city was Mayan. He seemed to agree.

I then told him of a study done by an electrical engineer on the Earth's rotation patterns. Mark Carlotto, author of *Before Atlantis* was surprised when I told him about the lost city of Cuba. He immediately asked for coordinates. After I informed him, he said it was much too deep to be explained by ice melting and the increased depth of the ocean. When I asked him whether it could be Mayan, he sent me the following map implying that there was *no land bridge* between the Yucatan and Cuba in the last 20,000 years which might explain it, either (see below).

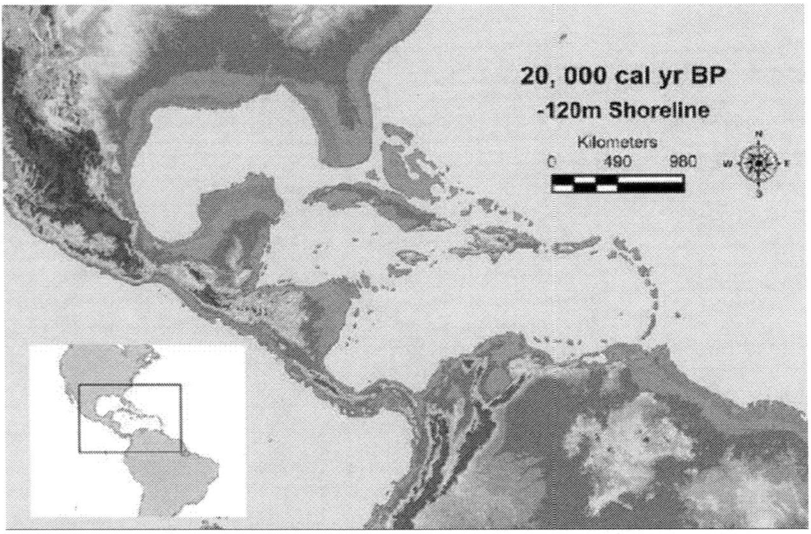

You can see in this photo, there is no land bridge connecting Yucatan with Cuba as far back as 20,000 years.

[103] George Erikson, Sunken ruins off Cuba may be linked to Maya, Graham Hancock website, September 2002,

https://grahamhancock.com/phorum/read.php?1,108197

The only explanation that makes any sense is that there must have been some kind of convulsion in tectonic plates which drove that structure 2,000 feet below sea level. And that observation leads us into the quixotic territory of Charles Hapgood.

Hapgood's Crustal displacement

When polar ice melts, it shifts the mass of the Earth toward the equator. Ice weighs less than water, and thus these shifts in mass might cause changes. The first change is that the rotation of the Earth slows. Curiously, even now with global warming, we are witnessing an ever-so-slight change in the Earth's rotation.[104]

> Whoops, we accidentally made the planet move: New research says human-caused climate change has accelerated the rate at which Earth's rotational axis changes. To get to the bottom of it, researchers from the Institute of Geographic Sciences and Natural Resources Research in China and the Technical University of Denmark pulled satellite data from NASA's Gravity Recovery and Climate Experiment (GRACE) spacecraft and studied this "true polar wander" phenomenon during a specific period of time in the 1990s. The scientists found moving masses of water have pushed Earth's axis eastward beyond even what existing climate models predicted. On the surface, this makes sense: Think about holding a basin of water and then moving it back and forth—the sloshing causes the water's weight to move all around. This is what's happening on a macro level.[105]

Secondly, there can also be changes in the 'wobble' of the Earth.[106] Imagine then the progression of changes during the end of the last ice age.

In 1966, Charles Hapgood proposed a theory that the Earth's pole had shifted abruptly some 11,000 years ago, and that there were gigantic movements of the Earth's crust as a consequence. Einstein was fascinated by Hapgood's work and wrote a preface to his book. Hapgood surmised that large parts of the Earth actually moved. He believed Antarctica moved as well as the Earth's crust shuddered.

---

104 Anthony Robledo, "We do not know how to cope" Earth spinning slower," USA Today, 3 April 2024.

https://www.usatoday.com/story/news/nation/2024/04/03/negative-leap-second-may-be-needed-study/73186366007/

105 Caroline Delbert, "Earths axis has secretely shifted," Popular Mechanics, 17 May 2021,

https://www.popularmechanics.com/science/environment/a36419756/earth-axis-shifted-climate-change/

106 "Are orbital changes causing global warming?" Earth@Home, 2007,https://earthathome.org/quick-faqs/are-orbital-changes-causing-global-warming/

I wrote about this subject in *Aliens and Man*. An excerpt from that text follows:

The maps of Pyri Reis and Oronteus Finaeus

Another mystery points also in the direction that *Homo sapiens* had deep knowledge of the world much earlier than supposed. In the early 1500s a Turkish navigator, Admiral Pyri Reis, drew maps of the world which he said were *based upon earlier maps* from which he worked. The map of Pyri Reis dates to 1513 AD.

What is remarkable is that the map describes rather accurately the continent of Antarctica. More peculiarly, it depicts mountain ranges and flowing rivers which correspond to images of the continent seismically surveyed *only in 1949*. A second map, the Oronteus Finaeus Map of 1531, showed Antarctica with non-glacial conditions in Queen Maud Land, Victoria Land, and the East coast of the Ross Sea. How could that be possible?

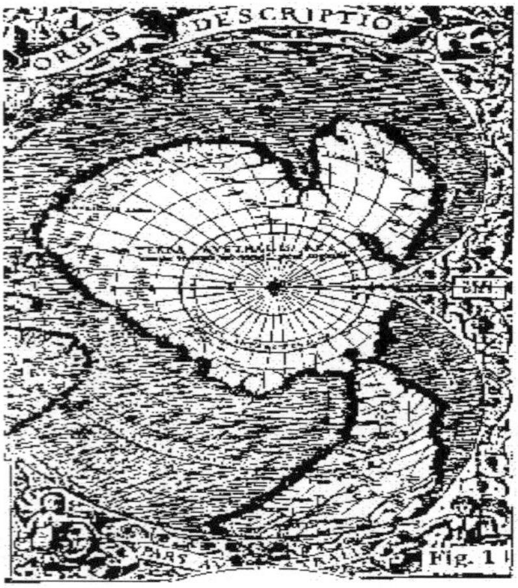

Fig. 1

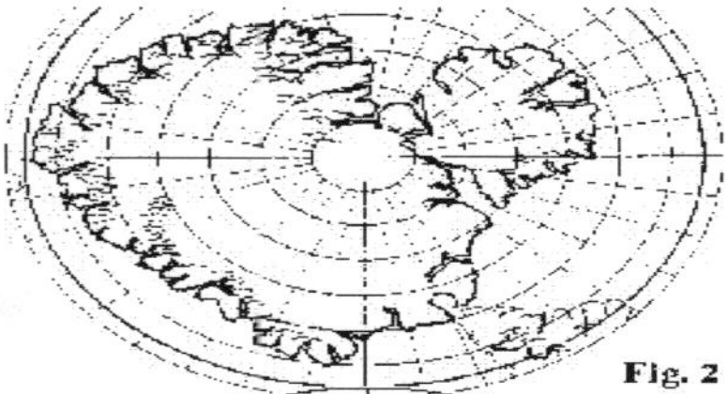
Fig. 2

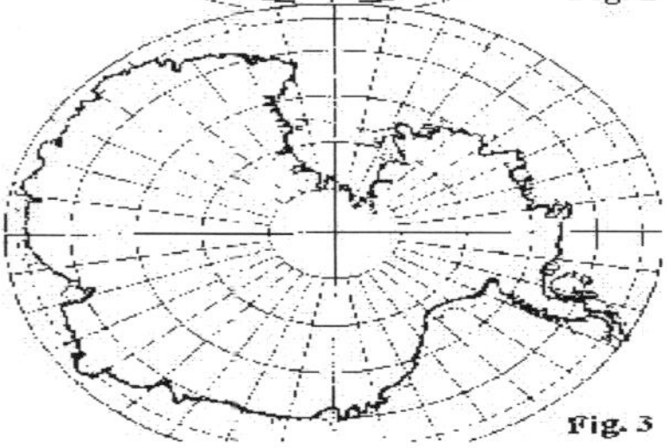
Fig. 3

1 - Oronteus Finaeus map of 1531, southern hemisphere

2 - The Oronteus Finaeus map redrawn on a modern polar projection

3 - A modern map of Antarctica drawn on a modern polar projection source: William Fuller, http://members.tripod.com/-Zomb/ANTARCTI.HTM

Antarctica was discovered by Captain Cook in 1799 but it was so shrouded in ice, he could not map its interior. From the Oronteus Finaeus map, *also copied from earlier sources*, ancient mariners were able to identify a mountain range that existed on the continent and also to have outlined its features when it was covered by an ice sheet miles thick. How was that possible? Modern projection techniques when applied to the map show it possesses uncanny accuracy. According to William Fuller:

'The Oronteus Fineaus map is more accurate than any map made anywhere up to the year 1800. Not only does this map show the entire continent of Antarctica, but also inland rivers and mountain ranges. These are no fanciful creations, the riverbeds and mountain ranges actually exist although today they are buried by a mile-thick cap. This fact was unknown until (the International Geophysical Year) when teams scientists did sonar tests of the ice which mapped the coastline beneath it. The modern maps match the old maps exactly.' [107]

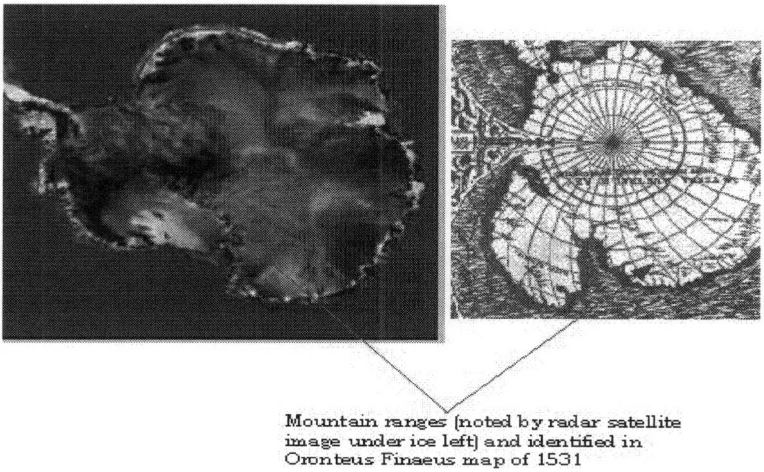

Mountain ranges noted by radar satellite image under ice (left) and identified in Oronteus Finaeus map of 1531

[107]"The Oronteus Finaeus map from 1531 depicts Antarctica 289 years before it was first sighted." Reddit, 2020

https://www.reddit.com/r/interestingasfuck/comments/i1zpw1/the_orontius_finaeus_map_from_1531_depicts/

Professor Charles Hapgood, author of *Maps of the Ancient Sea Kings,*[108] says the only way to know that mountain ranges existed on Antarctica was to have seen or witnessed them *before* their glaciation...The only problem with that conjecture is that there were no recorded civilizations then. Bit of a problem. Sumer and Egypt did not appear until the third millennium.

Longitude, we thought, was mastered in the 1720s, but these maps seem to betray knowledge of longitude and the capacity to do spherical trigonometry, the mathematics involved in computing projections on a globe."[109] One Russian scientist opines that:

....[B]etween the 5th and 10th millennium BC, there was a civilization on Earth that possessed great knowledge in the field of navigation, cartography, and astronomy, which was no less advanced than that of the 18th century.[110]

Obviously this is only circumstantial evidence that points to some lost and advanced civilizations that seem to have left traces. It is possible that some kind of advanced society existed prior to this time, antedating our knowledge of our own civilization. Skilled in mathematics, language, cartography, and astronomy, and through some unknown process of diffusion, it managed to spread its influences to the Egyptians and the Chinese as well as to have mapped Antarctica.

A good example is mythic Atlantis. That is a reasonable supposition. It does offer to explain away a lot of this data. But just how speculative and circumstantial is our data, we have to ask ourselves.

An Update on Oronteus Fineas

That excerpt comes from that book that I wrote two decades ago. There is an important update. I enlarged the Oronteus Fineas map below, and turned it upside down so you can see that the map clearly identifies rivers and tributaries running into the Ross Sea:

[108] Note also that the Oronteus Finaeus maps were originally studied by Dr. Charles Hapgood and Dr. Richard Strachan of MIT.Although Hapgood was treated by disdain by members of his profession, Albert Einstein found his work fascinating.

[109] Graham Hancock, *Fingerprints of the Gods*, New York: Three Rivers Press, p. 30.

[110] As an illustration of the vast treasures of knowledge lost to antiquity, it is said that the Egyptian Library in Alexandria in 323 BC possessed ten times the number of volumes that the entirety of Europe possessed in 1450AD. The Library in Alexandria was burned in 46BC, again in 391 AD and finished off by Arab legions in 640AD.

We know this entire area is covered by ice today (photo below) so the question is, at what time would anyone have known about tributaries running into the Ross Sea?

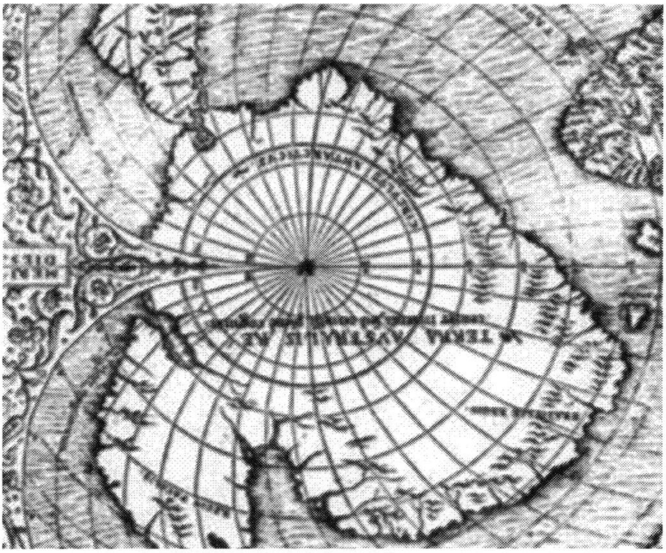

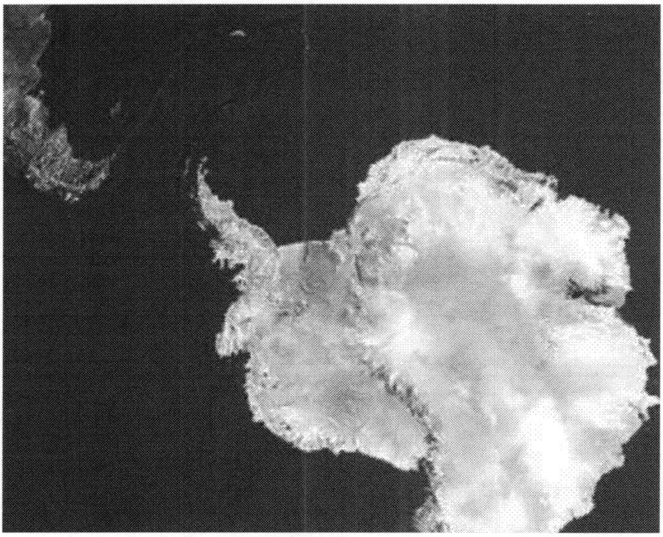

Satellite photo of Antarctica today. You can see the Ross Sea, the mountains, and tributaries seen on the Oronteus map are covered by ice.

Tributaries running into the Ross Sea

When was this part of Antarctica free of ice so that a map might be made?

A recent study published in *Nature*, with 20 authors, suggested that a period of melting began about 9,000 years ago.[111] A second study done in 2020, this time with a dozen Japanese authors, confirms that the melting began between 9000 and 8000 years ago.[112] Yet another pegs the melting around the Ross Sea to be about 7000-8000 years ago.[113]

Now what is curious is that western society was taught that Antarctica was first discovered in 1799. But someone appears to have mapped its coastline between 8000-9000 years ago?

Couldn't be Egypt—Egypt began only 5,124 years ago.

Couldn't be Sumeria —Sumeria began 6,524 years ago.

Couldn't be China—China began 2,400 years ago.

Who could it have been? Best guess: Atlantis!

A curious footnote

There is a strange paragraph in the Mayan book known as the *Troano Codex*. It describes a great catastrophe of the Earth's crust:

> The year six Kan, on the eleventh Muliic, in the month Za, there occurred terrible earthquakes which continued without intermission until the thirteenth Chuen. The country of the hills of mud, the land of Mu, was sacrificed. Being twice upheaved it suddenly disappeared during the night, the basin being continually shaken by volcanic forces. Being confined, these caused the land to sink and rise several times and in various place. At last the surface gave way, and the ten countries were torn asunder and scattered in fragments; unable to withstand the force of the seismic convulsions, they sank with their sixty-four millions of inhabitants, eight thousand and sixty years before the writing of this book.[114]

Since the book was written around 900 AD, it would put this cataclysm at 7160 BC. Thus, Mayan mythology seems to confirm what we have been alluding to in this book. And perhaps the Cuban lost city was sucked into the abyss due to the same set of processes.

---

[111] Yusuke Suganuma, et. al., "Regional sea level high stand triggered Holocene ice sheet thinning across coastal Dronning Maul Land," Communications and Earth Environment 9 Nov 2022 MaudHolocenehttps://www.nature.com/articles/s43247-022-00599-z

[112] Yosuke Yokoyama, et. al., "Widespread collapse of the Ross Ice Shelf," 16 Feb, 2016 PNAS,https://www.pnas.org/doi/full/10.1073/pnas.1516908113

[113] Brenda Hall and George Denton, "Deglacial chronology of the Western Ross Sea, Antarctic."Jounal of the United States, Review1996

https://www.nsf.gov/geo/opp/antarct/ajus/nsf9828/9828html/b16.htm

[114] Cited in *Before Atlantis*, ibid, p. 124.

More on Hapgood's theory

Hapgood believed there were major shifts in continents, crustal displacement of land masses as a consequence of melting ice and changes in planetary rotation. He proposed that the magnetic North pole shifted at least three times resulting in changes in the Earth's crust.[115]

The Hudson Bay position was thought to have occurred close to 12,000 BC. If true, then Antarctica might have been "several thousand miles north of is present position and would have had a temperate climate."[116] Carlotto in his text shows that several ancient sites are oriented toward the Hudson Bay position including Teotihuacan in Mexico.[117]

If correct, then we may have an entirely different hypothetical location for Atlantis. But we must also remain a bit aloof from this. Many in the geological community believe crustal movements are much slower and that Hapgood's theory is highly questionable. [118] [119]

A note about Immanuel Velikovsky

Velikovsky was a psychoanalyst who was one of the first to write about "catastrophism." He became interested in ancient history and from a close reading of mythology, he pegged a set of ancient catastrophes to have occurred about 3,500 years ago in the 15th century BC.

But a recent study of a funerary site in Egypt shows that the Egyptians rather accurately identified the solar cycle with an uncanny exactness 3,800 years ago.

> Additionally, the tomb "can be considered the oldest private funeral project in which a perfect intersolstice alignment is achieved, showing the presence of individuals with enormously sophisticated religious and technical training in provincial courts such as Elephantine," researchers wrote in the study.[120]

[115]"Earth Crust Displacement," https://www.bibliotecapleyades.net/sitchin/esp_sitchin_3g.htm

[116]Carlotto, ibid., p. 20

[117]Mark Carlotto, "A new model to explain the alignment of certain ancient sites," June 2020, Research Gate https://www.researchgate.net/publication/342255402_A_New_Model_to_Explain_the_Alignment_of_Certain_Ancient_Sites

[118] *Brass, Michael (July–August 2002).* "Tracing Graham Hancock's Shifting Cataclysm". Skeptical Inquirer. *26 (4). Archived from* the original *on 2010-04-26.* Retrieved 2010-04-19.

[119]*Tarduno, John A.; Smirnova, Alexei V. (January 15, 2001). "Stability of the Earth with respect to the spin axis for the last 130 million years".* Earth and Planetary Science Letters. *184 (2): 549–553.* Bibcode:2001E&PSL.184..549T. doi:10.1016/S0012-821X(00)00348-4.

[120]Brendan Racias, "Newfound tomb reveals ancient Egyptians had extensive knowledge of the solar cycle," The Island Packet, Dec 13, 2022, https://www.islandpacket.com/news/nation-world/world/article269948887.html

That effectively means that if the Egyptians precisely calculated the solar solstices, there was no crustal shift of movement in the Earth at least for 3,800 years. That fact certainly puts a damper both on Velikovsky and (though not entirely) on the Hapgood-Antarctica crustal movement hypothesis. The best evidence for crustal displacement currently seems to be the various alignments with the Hudson Bay pole as suggested by Mark Carlotto.

But certainly much more research needs to be done on this subject before we allow it to leave the hypothetical stage. So far, our evidence is just both hypothetical and speculative.

Gobekli Tepi

Our whole interest in Atlantis was to search for human culture and civilization that might have existed during the time of the ice age and which may have disappeared as a consequence of the cataclysms beginning in 12,000 BC, and extinction events that unfolded as a consequence.

One more sign of human cultural activity that may have existed before the Younger Dryas and which was wiped out with it is a place called Gobekli Tepi, and the peoples that may have built it, the Natufian culture.

Gobekli Tepi, located in Turkey, has been the object of scrutiny for some time. It defies the notion that hunter-gatherers were roaming about and that civilization, architecture and technological achievements didn't come until much later.

These few photos of Gobekli Tepe show its architecture and art. [121] Graham Hancock says the site has been "reliably dated" 11,600 years ago, just about the close of the Younger Dryas.

[121] Stephen Geist, Gobekli Tepe—an ancient anomaly inTurkey, Medium, 5 Aug 2023, https://medium.com/@stphngeist/g%C3%B6bekli-tepe-an-ancient-anomaly-in-turkey-6482ee23fe0d

That makes it 6,000 years earlier than Stonehenge and 7,000 years earlier than the Great Pyramid.[122]

Some archeological radiocarbon studies make it out to be younger and prefer 9,000 years before the present rather than 11,300.[123]

Be that as it may, this culture had some sense of design, an ability to shape stones, and one would be hard-pressed to call this anything but a mark of civilization. As Hancock surmises:

> [T]he complex design created by Gobekli Tepe's inhabitants, which transcends the abilities of any known civilization 11,600 years ago — right at the end of our cold snap, the Younger Dryas. To be able to build Gobekli Tepe, its creators must have gone through a "technology transfer," he adds, which means they learned to build the site from someone else. Someone older. The transfer didn't begin with Gobekli Tepe — which is itself 7,000 years older than Stonehenge. It didn't even begin in the Neolithic. It began millennia earlier with Late Epipalaeolithic cultures, one of which has, since the 1920s, been referred to as Natufian," Hancock argues. He points out that we have no idea what the Natufians called themselves, nor whether they were one or multiple cultures."[124]

Gobekli-Tepe artistic carving

[122] Wikipedia Stonehenge, https://en.wikipedia.org/wiki/Stonehenge

[123] Dimitrios Dendrinos, "More on Carbon 14 evidence and Bobleki Tepe,' Acadmia.edu, 2021 https://www.academia.edu/49493094/MORE_ON_CARBON_14_EVIDENCE_AND_GOBEKLI_TEPE

[124] Kay Smythe, "Graham Hancock Reveals Huge Update on Mysterious Ancient Archeological Site," The Daily Caller, 16 April 2024, https://dailycaller.com/2024/04/16/graham-hancock-ancient-apocalypse-big-archaeology-turkey-younger-dryas-natufian-israel/

Modern archeology holds that the first settlement of human beings, something resembling a city, was Jericho, considered the oldest settlement in the world and dated to 10,000 years ago. [125] But Gobekli Tepe happens to date back more than a millennium and a half *before* Jericho.

Even more startling is that if Gobekli Tepe is 11,600 years old, it would be contemporaneous with Atlantis—that is, if Atlantis existed! Unlike Atlantis, though, Golbelki Tepe does indeed exist. "Göbekli Tepe is one of the first manifestations of human-made monumental architecture," says UNESCO [126]

Archeologists are still working Gobekli Tepe, and they believe it will take decades more to decipher the kind of civilization that existed there. But there are some hypotheses about the technology they used. They had rather precision stone tools for cutting.

> We see the products of printing microblade technology... The printing microblades we are working on are 2-5 mm. They are tiny cutters. These are precision tools used in the finest works by arranging them on bone material. Stone tool production technology has obviously been moved from east to west from about 7,000 kilometers.[127]

The Natufians?

Hancock mentions a culture I never heard of, the mysterious Natufians. This group, he surmises, may have 'taught' the masons of Gobekli Tepe how to work their stones. Interestingly, the Natufians existed *prior* to the Younger Dryas and are dated at 15,000 years ago.[128]

They were already a sedentary agricultural population and considered the first to develop bread, cultivate barley, build stone houses, and rudimentary art. Evidence of bread goes to the time *before* the Younger Dryas.

The map of the Natufians places them quite near Gobekli Tepe, slightly south.

The tools of the Natufians were "bedrock mortars, portable mortars, bowls of various types, cup holes, mullers, and pestles, occur in large numbers in base-camp sites, but are not as abundant in the more ephemerally occupied camps." [129]

As I examine this literature, it does not convincingly appear to be an explanation for Gobekli Tepe. The Natufians did not have any structures

[125] "Jericho, An ancient city revealed," Oriental Museum, Durham University. No date, https://stories.durham.ac.uk/jericho/

[126] "Gobekli Tepe" Unesco 14 Jul, 2018 https://whc.unesco.org/en/list/1572/

[127] "Gobekli Tepe reveals stone tool technology," Daily Sabah, 27 June 2022, https://www.dailysabah.com/life/history/gobeklitepes-latest-mystery-reveals-stone-tool-technology

[128] Natufian culture, Wikipedia https://en.wikipedia.org/wiki/Natufian_culture

[129] Ofer Bar-Yosef,"The Natufian Culture in the Levant, Threshold to the Origins of Agriculture" Columbia University, Evolutionary Anthropology,1998

resembling Gobekli Tepe, and the suggestion that they could have been the authors or influencers of Gobekli Tepe seems unjustified.

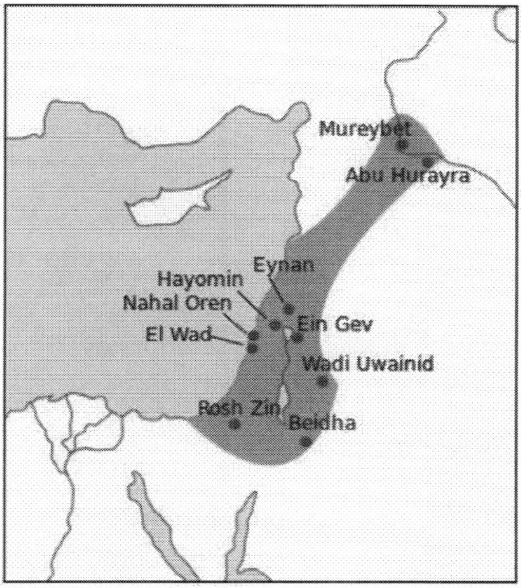

Natufian sites

Graham Hancock, it is said, believes Gobekli Tepe was constructed by the survivors of Atlantis. At first look, that might be quite possible. Clearly, the Natufians do not seem up to the task.[130]

There is another site nearby called Wadi Hammeh in Jordan that dates to at least 12,000 years, but, again, it does boast of any mathematical or structural evidence of an advanced culture.[131]

Many archeologists call Graham Hancock a pseudo archeologist and pooh-pooh his claims, but there is a remarkable assemblage of data on Gobekli Tepe which shows a mathematical sophistication that doesn't seem to exist anywhere else at this time. [132]

So, let us continue looking for some real advanced civilization that may have existed before the Younger Dryas.

[130] J Laperle "Pseudo Archeological Claims," 15 March 2022, https://blogs.uoregon.edu/gobeklitepe/

[131] Shelby White, "Waki Hammeh 27," The President and Fellows of Harvard College, 2013. https://whitelevy.fas.harvard.edu/wadi-hammeh-27-early-natufian-settlement-pella-jordan

[132] Haklay, & Gopher, A. (2020). Geometry and Architectural Planning at Gobekli Tepe, Turkey. Cambridge Archaeological Journal, 30(2), 343–357. https://doi.org/10.1017/S0959774319000660

So far, everything is still hypothetical.

The Yonaguni Monument, Japan

About 266 km from Taiwan lies the small island of Yonaguni, which houses an incredible structure offshore called the Yonaguni Monument.[133] It is 85 feet underwater, so it certainly would qualify as a monument that might have been submerged after the deluge and the flooding of the Younger Dryas. It appears to be a human-made construction. Following are a number of photos a diagrams of the structure:

Wikipedia makes sure to cite three different authors who say these are natural processes, and that it is not human made at all. Those alleging these might be actual human constructions are immediately labeled 'pseudoarche-ologists.' [134] [One can legitimately wonder at the unbiased nature of Wiki-pedia.]

At any rate, there doesn't seem to be any consensus at this writing as to whether it was natural or a human-made construction.

Some observations: The island is quite tiny. Today the population is only about 1,600. It is Japanese but quite far from Japan proper, about 500 km from Okinawa. The structure is about 45,000 square meters.

[133] Frank Engel, "Yonaguni Monument," July 2020

https://www.google.com/maps/place/Yonaguni+Monument/@24.4354386,123.0112256,3a,75y/data=!3m8!1e2!3m6!1sAF1QipMRwtyJbQTZBh8GWeM8b3MUX2XAhnSiTS-XshJT!2e10!3e12!6shttps:%2F%2Flh5.googleusercontent.com%2Fp%2FAF1QipMRwtyJbQTZBh8GWeM8b3MUX2XAhnSiTS-XshJT%3Dw203-h115-k-no!7i2328!8i1325!4m7!3m6!1s0x34674dc801679 1b5:0x335ca95615e2e2c1!8m2!3d24 4354386!4d123.0112256!10e5!16s%2Fm%2F04y5lvb?entry=ttu

[134] Wikipedia, Yonaguni Monument. https://en.wikipedia.org/wiki/Yonaguni_Monument#:~:text=%E2%80%A2%20Depth%2026,m%20%2885%20ft%29

One interesting deep-dive archeological expedition reported findings I found interesting. They believe the structure was antediluvian and created sometime before 10,000 BC. In the lengthy article they describe children in 1967 diving who found a carved head near the monument which they describe as follows:

> The carved stone head found by these children, actually explorers of Yonaguni Diving Team (YDT) found for them in the underwater area off the tip Araka Wabana on the south coast of the island, YS-CR1 is in itself a clear anthropological and precise description of archaic people, which leaves no room for conjecture; craniofacial morphology with rugged, narrow and elongated skull with powerful facial bones...[135]

This pdf has an irritating mistranslation of BC and AD, and it causes one to wonder about its overall accuracy. This stone head was not found near the Yonaguni monument, but it was underwater.

But also reported in this expedition was this carving or tablet which boggles the mind:

Note that the sandstone tablet actually pictures a step-like pyramid in the center. I did not see any ruins on Yonaguni which might resemble this, so it seems certainly possible the tablet is portraying the underwater monument, perhaps as a step-pyramid.

[135]Prieto, M.A., 1985 Archeology of Yonaguni, pdf. P. 74

I was so amazed by these findings—and so certain that Yonaguni was made by a human civilization before the Younger Dryas—that I had to calm myself down and try to double-check. I wrote to Dr. Miguel Prieto, the author and leader of the diving team, to verify that I was understanding this evidence properly.

Head of carved stone (YS-CR1) with the Ameridian distinctive features that describes the authors of the work (YS-I)

PRIETO, M.A.(1985) UNDERWATER ARCHAEOLOGY OF YONAGUNI JIMA YAEYAMAISLANDS. JAPAN. (VOL. I)

YF-I fragment of a tablet carved sandstone found in YS-I by the team of Yonaguni Diving Team (Underside). The inscription refers to an ancient town of the Archaic Navigators, called Aat-Aragua (Taragua) unidentified. It may be the called "Aat Re Mu" (Mu sunk Country District) after Event Cariaco. But this specifically relates back to a family identified with your name, to a mountain temple and an observatory (Cue).

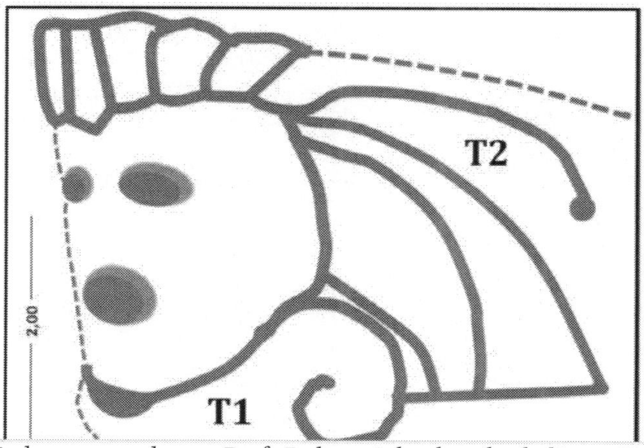

I also wrote also to Prof. Robert Schoch, who believes Yoganumi was carved by natural processes. He still believes that, even after I presented him with the evidence of Dr. Prieto.

Dr. Prieto replied to me and affirmed that the sandstone petroglyph is dated at the same time as the Yoganumi monument.[136]

He also underscored that the Toronoico statue is convincing, too, and at a depth of 40 meters.

[136]personal email.

The Jomon culture

A further line of reasoning here is that the Japanese Jomon culture existed perhaps as early as 14,000 BC. Pottery and other excavations prove they had some degree of sophistication. Here is an excerpt of that study:

> The oldest of these is the so-called Tōryūmon doki (Figure 1), found at the site of Ōdai Yamamoto I (in Aomori Prefecture), dated to 16,000BP, thus around 14,000 BC. Leaving aside the strong doubts about the robustness of the dating performed, this would make Japanese pottery the oldest in the world. In any case, according to this school of thought, the beginning of the Jōmon Period would be dated to 14,000 BC. Apart from the uncertainty over the proposed dating, it must be said that the fragments found of such pottery are so few in number and limited to a single location, that they cannot be explanatory of a large-scale evolution [8]. The second theory would instead place the beginning of the period at 14,500 BP (ca. 12,500 BC), from the dating of the Ryūkisenmon doki (Figure 2) identified as the first pottery type to spread across much of the country, northwards [7]. The third, finally, would place the 'starting point' of the Jōmon Period at 11,500 BP (ca. 9,500 BC), the period to which the Yoriitomon doki (Figure 3) is dated, associated with other elements that highlighted the development, in this period, of gathering, hunting and fishing.[137]

So perhaps the Yoganumi Monument was constructed during the Jomon period.

But before our speculations get any wilder, remember that a rather sound academic peer-reviewed study dated the monument at best at 10,000 years. [138] This is *after* the Younger Dryas, and although certainly old, current research doesn't really support the suggestion that an advanced civilization existed prior to the beginning of the Younger Dryas at 14,000 BC.

And that fundamentally is what this text is trying to investigate in Part II.

In other words, close but no cigar — at least not until future research shows the Jomon people possessed mathematics, written language, and quite a few important other cultural artifacts beyond pottery and hunting and gathering.

---

[137] Daniele Petrella, "The Japanese Neolithic: The sea at the Origins of the Jomon Culture,"Anthropology Open Access, 2023.https://www.academia.edu/100549855/The_Japanese_Neolithic_The_Sea_at_the_Origins_of_the_J%C5%8Dmon_Culture_and_the_Iriae_Archaeological_Excavations_on_Tsushima_Island?email_work_card=thumbnail

[138] Kimura Misaaki, et. al., "Age determination of the submarine ruins of Yonaguni Island Okinawa by means of surface irradiation method," ETDE Web, 01 Sept 2003, https://www.osti.gov/etdeweb/biblio/20422305

fantasy idea of what Gunung Padang looked like 20,000 years ago

Fantasy rendition of what Gunung Padang looked like 20,000 years ago

The pyramid of Gunung Padang

In Indonesia there is a controversial structure called Gunung Padang. The site is 3,000 feet above sea level, covers 72 acres, and has a number of terraces which become smaller with each elevation, thus resembling a step-like pyramid. It was first reported in 1890. Some consider it to be an ancient pyramidal structure dating to before the Younger Dryas, but Wikipedia describes that hypothesis less generously:

> Fringe hypothesis: Danny Hilman Natawidjaja, an Indonesian geologist, has claimed that the site had been built as a giant pyramid 9,000 to 20,000 years ago, implying the existence of an otherwise unknown advanced ancient civilization. Natawidjaja's analysis has been questioned by other scientists. Vulcanologist Sutikno Bronto concluded that Gunung Padang is the neck of an ancient volcano and not an artificially created pyramid. Archaeologist Víctor Pérez has described Natawidjaja's conclusions as pseudoarcheology. [139]

Evidence it was made by human hands

Let us at least examine the hypothesis before rejecting it. Here is a statement of its basic premise:

[139] Wikipedia, "Gunung Padang," https://en.wikipedia.org/wiki/Gunung_Padang

> For this new study, the research team...studied the structure using seismic tomography, electrical resistivity tomography and ground-penetrating radar. They also drilled down into the hill and collected core samples that allowed them to use radiocarbon dating techniques to learn the ages of the layers that make up the hill.
>
> [T]he research team found what they describe as clear evidence showing that the Gunung Padang was made mostly by human hands..... the structure was built in stages, thousands of years apart. . .the older parts of the structure were made sometime between 25,000 and 14,000 years ago, making it the oldest known pyramid in the world today....
>
> The researchers found evidence of several efforts that together over time, added up to a completed structure. The first consisted of sculpted lava—where builders had carved shapes onto the top of a small, dead volcano. Then, several thousand years later, sometime between 7900 to 6100 BCE, another group added a layer of bricks and rock columns. Some unknown time later, another group added a dirt layer to part of the hill, covering some of the earlier work. Then sometime between 2000 and 1100 BCE yet another group added more topsoil, stone terracing, and other elements.
>
> The research team has also found some evidence suggesting there might be some hollow parts inside the structure, suggesting possible hidden chambers. They plan to drill down to them and then lower a camera to see what might be in these areas.[140]

Fine, but rather than carving stone boulders and placing them into precise positions, it looks like they carved lava boulders that were already there and placed them into an appropriate alignment. It doesn't quite seem the same as the Great Pyramid or, for that matter, Gobekli Tepe where stones appear to have been quarried from many miles away.

Academic archeology reacts

Traditional archeology takes issue with all of this, of course, and dates the site to about 1500 years ago. Here is their rebuttal:

> Their insistence that the hill was "constructed" just doesn't make sense. Particularly given the sheer amount of columnar jointing found associated with volcanic activity. This sort of rock formation is created when lava cools and it continues to extrude from the central magma

[140] Bob Yirka, "New evidence strongly suggests Indonesia's Gunung Padang is oldest known pyramid," Phys.Org, 6, Nov, 2023.https://phys.org/news/2023-11-evidence-strongly-indonesia-gunung-padang.html

chamber, either through vertical or near vertical shafts or even horizontal ones.

Geologists looking at volcanoes in the vicinity, including Gunung Padang, describe the hill as being the youngest plug of andesitic basalt lava with columnar structure in the crater of an ancient volcano....This volcano would have been active until around 32 million years ago.[141]

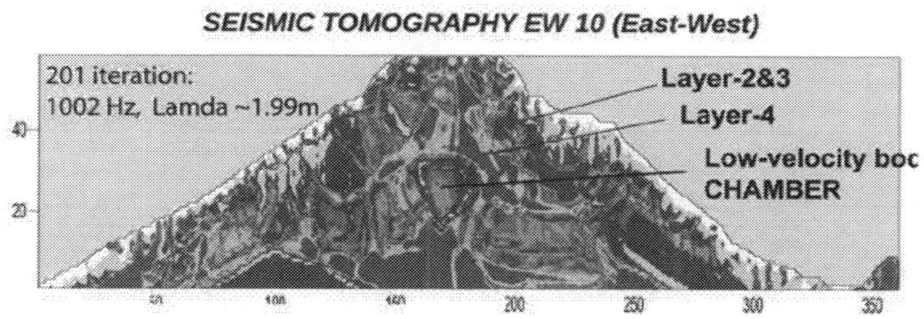

As of this writing, it appears the establishment is winning. A scientific article was recently retracted, and now it looks like the 20,000–27,000-year-old structure is not likely to pass the smell test.[142]

The New York Times

Study About Purported Ancient 'Pyramid' in Indonesia Is Retracted

There is a reluctance in the Indonesian government to drill to an area which might resolve this question. Yet there is a cavern or chamber 50 feet down, never seen before, and over 36 scientists have attempted to *prevent* the exploration. All quite unusual. Ground-penetrating radar suggests such a chamber exists.[143] Perhaps the chamber will one day cough up its secrets, but, for now, one cannot assume that a technologically advanced society built

[141] Carl Feagans, "Gunung Padang: what archeology really says," Archeology Review, 28 august, 2023, Pshttps://ahotcupofjoe.net/2023/08/gunung-padang-what-archaeology-really-says/

[142] "Study about ancient pyramid is retracted," New York Times , 03 March 2020. https://www.nytimes.com/2024/03/20/science/indonesia-oldest-pyramid-gunung-padang.html

[143] Mark Brockman,"Has Graham Hancock been vindicated?" 17 Nov 2023, https://www.YouTube.com/watch?v=Zal63OSlwDI

this structure prior to the Younger Dryas until much more data comes in to support that proposition.

Summing up

From a review of the literature which we have attempted, I think the conclusion is disappointing. If you were hoping to find a technologically advanced, robust society that was antediluvian, and by that I mean, an advanced human society prior to 12,000 BC, I think we are fundamentally discouraged.

The best evidence, in my view, was Gobekli Tepe. It is real. It has a sophistication that implies cognitive intelligence, mathematics, and some engineering skill. But whether it existed *prior* to 12,000 BC is indeterminate. Radiocarbon dates take us only out to about 11,000 years ago.[144] Secondly, while it has been lionized as a site of great technical prowess, there is little evidence it had any sort of stunning existence prior to 12,000 BC and prior to the Younger Dryas. What's more, the tools they used do not seem very sophisticated at all—quite stone-age frankly, and nothing remarkably advanced from other prehistoric peoples using stone implements.

The next runner-up in my view is the Yonaguni Monument. It has striking qualities, especially with the stonework seemingly "crafted" and systematically organized. But it dates to only about 10,000 years ago, and only if you are being generous and ignoring the contrary dating can you place it before the Younger Dryas. [145]

The lost city of Cuba is certainly at a depth that might make it qualify for antediluvian status having been buried deep beneath the sea, but frankly, there is insufficient data to draw any real conclusions today. Far more research is needed before we engender significant enthusiasm.

Frankly, the most intriguing specimen on our list is Atlantis. The only trouble is, we can't find anything to prove it actually existed. It would be truly antediluvian, truly advanced, truly technologically sophisticated, but all we have is circumstantial evidence that it existed: mythological tales, the Bimini Road, perhaps the lost city of Cuba, and the strikingly similar in stonework in Egypt and Peru. That technology certainly would be hard indeed to spread directly from Egypt to Peru, over 15,000 miles, and much more easily disseminated from Atlantis which is almost equidistant from these two locales.

[144] Oliver Dietrich, "Radiocarbon dating the first temples of mankind," Research Gate, Jan. 2011 https://www.researchgate.net/publication/258182967_Radiocarbon_dating_the_first_temples_of_mankind_Comments_on_14C-Dates_from_Gobekli_Tepe

[145] Kristine De Abreau, "Exploration Mysteries, Yonaguni Monument" Explorers Web, 27 April 2023, https://explorersweb.com/exploration-mysteries-yonaguni-monument/

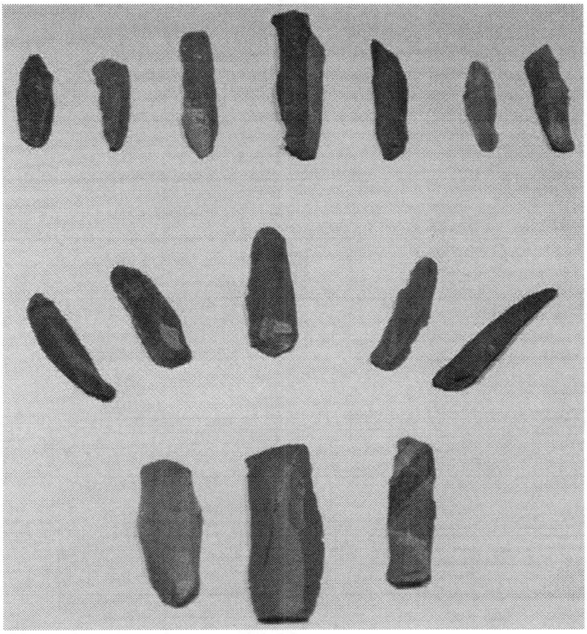

Stone tools used at Gobekli-Tepe: source: https://bikeclassical.blogspot.com/2014/11/gobekli-tepe-beginning-of-history.html

Further, the map of Oronteus Fineas does imply that some kind of advanced aquatic culture existed on this planet that successfully mapped Antarctica when the Ross Sea was ice-free, and that takes us back to roughly to the time Plato said Atlantis existed, namely 9600 years ago.

But, to be honest with ourselves, all we have is circumstantial evidence there ever was an Atlantis.

I'm not dismissing Edgar Cayce, either. I think it was positively amazing he mentioned a little 8-mile long Bimini island in the context of Atlantis, three decades before anyone discovered the Bimini Road.

But the order of the day it seems is that we have to 'curb our enthusiasm,' and begrudgingly admit we have not discovered any hard evidence of an advanced human community that existed prior to the Younger Dryas. We have not crossed the boundary of hypothetical and speculative with any data we have uncovered.

Homo Sapiens was here for 196,000 years, yes, but the first evidence of genuine human cognitive skills and advanced culture is really Gobekli Tepe which is currently dated 3,000 years after the Younger Dryas.

Perhaps one day we can say it was Atlantis, but today we can't. All we can say is there is insufficient data to support it.

Bias, distortion, and resistance:

While I have an abiding respect for science and formulating our conclusions on solid foundations, there is a peculiar stream of bias running through the social sciences which has not yet seemed to infect the harder sciences. For example, Graham Hancock is reviled in academic anthropology and archeology. And any professor who talks about Atlantis is seen as a lunatic and certainly not going to get tenure in the hallowed halls of academia. That much we know. There is a strong level of intolerance for many ideas and concepts that seek to gain some attention in these disciples.

But in physics, there is a lively debate going on over multiple controversies. Dark energy has never been directly observed. The same holds to dark matter. It is just as illusive as Atlantis; no direct evidence at all, only circumstantial portents. And yet there are job openings in astrophysics for people specializing in the study of dark energy, dark matter, quantum gravity, the QCD vacuum, "strange matter" even though many of these ideas have yet to show any evidence they even exist.[146]

Recently, the Webb telescope showed galaxies that are just not supposed to be there. They contradict the theory of cosmic expansion, very fundamental concepts in astrophysics, and the very origin of the universe.[147] But the contradictory data is *embraced and* discussed. A major conference is currently being held to try to sort out all these contradictions.[148] It is open, transparent, and no one is being canceled, censured, or fired for holding such hypothetical ideas.

But in the social sciences, it is far different. I wrote to a professor of anthropology, a card-carrying member of the establishment, and I told him I was going to write this text and that I didn't want it to be laced with conspiracy literature. I asked if he could please recommend a book that talked about human civilizations *before* the Younger Dryas. He suggested *The Dawn of Everything: a new history of Humanity* by David Graeber and David Wengrow.

[146] Wikipedia, List of unsolved problems in physics, https://en.wikipedia.org/wiki/List_of_unsolved_problems_in_physics

[147] Teresa Nowakowki , "Webb telescope finds evidence of massive galaxies that defy theories of the early universe," Smithsonian, 24 Feb, 2023, https://www.smithsonianmag.com/smart-news/webb-telescope-finds-evidence-of-massive-galaxies-that-defy-theories-of-the-early-universe-180981689/

[148] Chris Evans, "Upcoming Science with the Hubb & Webb conference,", Science Announcements, 7 march 2024 https://esahubble.org/forscientists/announcements/sci24002/

I read it and commented that there was no discussion of Atlantis, the lost city of Cuba, the Bimini Road, Younaguni, Gunung Pedang, or even of Gobekli Tepe. Nothing more than the hunter-gatherer Clovis people seemed to exist prior to the Younger Dryas. When I asked why all this was left out, he replied, with some apparent irritation, as if to say why should the book talk about things that don't exist.

I believe there is enough circumstantial evidence to suggest that a search for Atlantis should be a consummately legitimate endeavor, and exploring the nature of the submerged structures of Cuba certainly deserves attention. Why be snooty about exploring such exciting hypotheses more thoroughly?

Graham Hancock only has a BA in sociology, and some of his conclusions may be on the cusp of speculation, but he is a grand explorer like an Indiana Jones of the twenty-first century. He dove over 200 times to examine the Yonaguni Monument. Why no applause at all from the academic community? His work is astonishing and absolutely worth taking seriously.

If Graham Hancock's work was in physics, there would be a large colloquium looking into these matters with zeal. No one would be "banished," no one frozen out of grant money. No one would be rejected for tenure for looking suggesting hypotheses or trying to examine such incredible mysteries. But the social sciences are so infected with conformity and a strange new 'woke' ideology, that genuine academic curiosity or "exploration" is hardly to be found.

I am a psychologist, another social science, and there are so many forbidden topics it is hard to keep track of them all. If you assert sex differences in behavior, or brain functioning, you are treading on very turbid waters. If you assert there are only two sexes, XX and XY, you are in for censure and sanction. [149] If you mention race in the wrong way, you can be fired.[150] If you "misgender" a student's name, you can be fired or required to attend "retraining" sessions to keep your job.[151] Failing to use the appropriate

[149]"Neuroscience professor removed from APA discussion after saying there are only two sexes"14 May, 2021, Newsweek, https://www.newsweek.com/neuroscience-professor-removed-apa-discussion-after-saying-there-are-only-two-genders-1591697

[150]Jack Stripling, "The professor is canceled, now what?" Washington Post, 21 June, 2021 https://www.washingtonpost.com/education/2023/06/21/college-professors-fired-cancel-culture/

[151]Cameron Abrams, Christian Biology Professor Fired," The Texan, 28 June 2023, https://thetexan.news/issues/social-issues-life-family/christian-biology-professor-fired-from-san-antonio-college-after-complaints-over-statements-on-religion-gender/article_f6ac6c68-de77-58b0-b772-8a1f2c56d9bb.html

pronouns can get you terminated. [152]A tenured professor can also be fired for tweeting too.[153]

Sociology is not far behind with a litany of fired or censured professors who violated the new ideology.[154]That is the current zeitgeist in the social sciences, and I believe it creates an atmosphere of caution and reticence to explore anything new, anything controversial, or anything not within proscribed academic party lines.

And that is certainly unfortunate for exploring any of the topics that we have explored here.

So, to be comfortable and uncontroversial and to keep our respective tenure track positions intact we must restate that all the extinctions of the megafauna noted during and after the Younger Dryas were not caused by any cosmic events at all, but that a gradual ending of the ice ages began through normal cycles, and human-hunter-gatherers were responsible for killing off all the mammoths, mastodons, saber-tooth tigers and the other 32 species during this period.

End of story.

Now you can go home, have dinner and not worry about job security any longer.

---

[152] Matt Lavietas, "Professor who wouldn't' use trans pronouns wins $400K settlement," NBC News. 19 April 2023,https://www.nbcnews.com/nbc-out/out-news/professor-wouldnt-use-trans-students-pronouns-wins-400k-settlement-rcna24989

[153] Alex Galbraith, "UCF professor fired following controversial tweets about Black privilege has job reinstated, "Orlando Weekly, 20 May, 2022, https://www.orlandoweekly.com/news/ucf-professor-fired-following-controversial-tweets-about-black-privilege-has-job-reinstated-31635697

[154] Peter Borilla, "Fired for trying to teach sociology," FIRE, 12 Aug 2015, https://www.thefire.org/news/fired-trying-teach-sociology-former-professor-takes-rowan-college-gloucester-county-court

PART THREE. A QUIETLY REBELLIOUS THEORY

In trying to sort out human origins, I'd like to present a consummately different approach. A few years ago, I wrote a book about UFOs and explored the evidence as best I could. I tried to be meticulous in such a contentious area and I sorted out ostensibly valid accounts from craziness. The result of that effort was a book entitled *The Roswell Foil: how UFO discoveries change how we see the world.*

In the process of writing that 600-page tome, I gathered 520 witnesses whom I seriously vetted. In other words, out of the thousands of witnesses to UFOs, 520 survived this scrutiny and were considered worthy of attention. Of these, 240 were "expert" witnesses, namely people who had a special knowledge or training. Among them were 24 generals, many government "insiders," and scores of pilots, navigators, cosmonauts, astronauts, astronomers, professors and majors and captains in the military.

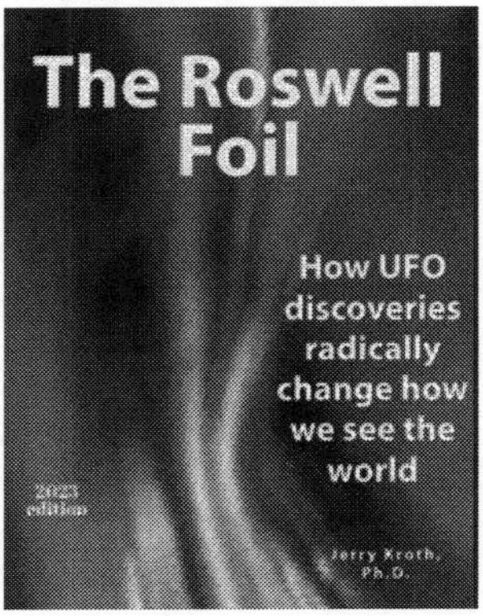

The conclusion of the first part of the book was clear. If there were a true, honest, transparent congressional hearing which allowed all these witnesses, their death-bed confessions and affidavits to be heard, there would be one conclusion: extraterrestrial visitations of Earth have happened and continue to happen to this day.

Why I am telling you this?

The last portion of that book examined a curious idea: "How would our lives change if we granted that extraterrestrials have interacted with humankind?" That knowledge, or consciousness, would represent an intellectual paroxysm larger than anything we have ever encountered: bigger than the revelations of Copernicus or Galileo, and far more earth-shattering than Einstein's theory of relativity.

Most of our academic disciplines would have to be rewritten. Our fundamental understanding of human origins, that is, anthropology, archeology, religion, and mythology all would come under revision. We would wonder if Zeus was really a real being, or if Quetzalcoatl perhaps wasn't just mythic. Did the gods actually mate with humans as mythology so obsessively seems to report, especially Greek mythology.

As such, I would like to present two chapters to you from that book which elucidates that hypothesis. It indeed may be relevant to the matters we have been discussing. We will skip the UFO part, but understanding the cultural implications and impact is what makes this undertaking relevant.

In other words, perhaps the Pyramids didn't come from Gobekli-Tepe or Atlantean 'influences.' Maybe they came from another kind of influencer—ancient aliens!

You may think he is going to take you for a long, crazy, pseudo-scientific ride, indeed; but remember one thing. In physics, scientists propose unusual, unverified, and entirely speculative ideas: parallel universes, multiple universes, even that this reality is a simulation—but they aren't laughed out of the room and ostracized.

I hope you will indulge these two chapters as one possible answer to the mysteries we have been exploring. Is it possible that the rapid increase in human knowledge, written language, and technology which only dates to the last 10,000 years, and which seems to only appear after the Younger Dryas, could have emerged from *outside* influences and manipulations. That is the contentious notion we shall examine next.

Why did it take *Homo sapiens* 194,500 years walking this planet before he created a written language? Why did 194,000 years pass before our species suddenly learned to domesticate pigs?

One answer to these myriad mysteries is that something rather miraculous happened to our species and very, very, recently. We may be the only beneficiaries of these manipulations, all coming in the last 10,000 years. These changes did not happen to Neanderthal who was here for 280,000 years. They didn't happen to Homo Erectus who roamed these plains for 1.5 million years, nor to Australopithecus whose tenure lasted 2 million years.

We will seek an answer in the next two chapters.

Pictographs and pyramids: an archeological journey

> Shrines! Shrines! Surely you don't believe in the gods.
>
> What's your argument? Where's your proof?
>
> —Aristophanes, 411 BC

There are many academic controversies when we look at archeology and the ancient alien hypothesis. First, we trip over Erich Van Daniken, who writes of ancient astronauts riding *Chariots of the Gods*. Van Daniken sold 44 million copies of his book, but his popularizing and failure to provide sources and references threw his work into disrepute in academic quarters.

Van Daniken claimed that the Nazca plains in Peru were landing areas for ancient spacecraft, but if extraterrestrials could manage to fly light-years across interstellar space, would they need special zones all mapped out for them to land?

Nazca lines in Peru as seen from the air

On the other hand, the cargo cult of the Caroline Islands had never seen an airplane until a man from California landed his Cessna there in 1941 and gave them various presents. After he left, they 'deified' this man and called him "John Frum." His name was John and he was *from* California, but that all got translated into a God who came from the sky who was named "John Frum."

As years went on, rumors of the god from the sky grew. Each year they built a bamboo plane in hopes John Frum would see it again and return with that precious second-coming cargo, how different is that than a primitive people making drawings on the Nazca plains so that space creatures would see their markings and return for a second visit? If the cargo cult built a mock aircraft and an airfield to *beckon* John Frum back from the heavens, are the Nazca figures similar attempts to charm star people back to Earth? This Nazca figure in the diagram here, commonly called "The Astronaut," is 32 meters in size and hard to see at all except from the air.

UFOs, Beings, and Artifacts

There is a ceiling painting found in a 3,000 year-old Temple near Cairo showing what looks remarkably like a helicopter. And there are Dogu figurines in Japan estimated at 2,500 years of age which show creatures in something akin to a spacesuit. [155]

[155] "The Abydos hieroglyph does not depict a helicopter," 1996, Bruckhttp://members.tripod.com/-a_u_r_a/abydos.html

One day a researcher was meandering through the Egyptian Museum of Antiquities. [156] He discovered an object previously thought to be a figurine dated to about 200 BC. It turned out to be a seven-inch wooden airplane made out of sycamore, which *could actually fly*.

> It possessed characteristics never found on birds, yet which is part of modern aircraft design. Dr. Messiha, a former model plane enthusiast, immediately recognized the aircraft's features and persuaded the Egyptian Ministry of Culture to investigate. Made of very light sycamore the craft weighs 0.5 oz. with straight and aerodynamically shaped wings, spanning about 7 inches. A separate slotted piece fits onto the tail precisely like the back tail wing on a modern plane.

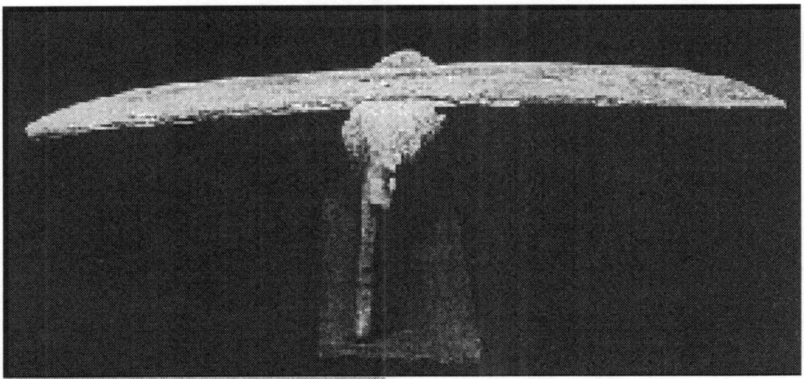

Egyptian sycamore glider; hieroglyph which looks like helicopter (above)

A full-scale version could have flown carrying heavy loads, but at low speeds, between 45 and 65 miles per hour. What is not known, however, is what the power source was. The model makes a perfect glider as it

[156] Feats and Wisdom of the Ancients: Time-Life, Alexandria, Virginia, 1990.

is. Even though over 2,000 years old, it will soar a considerable distance with only a slight jerk of the hand. Fully restored balsa replicas travel even farther.[157]

Mayan sarcophagus thought by some to depict Pakal sitting at the controls of a mysterious engine whose fiery exhaust emerges at the bottom of his chair.

[157] Joseph Jochmans, "Top 10 Out-of-Place Artifacts" Atlantis Rising, #5,

A Mayan sarcophagus allegedly portrays Pakal in a flying machine. The carving was found in Palenque and can be seen in the Mexico's National Museum of Anthropology. [158]

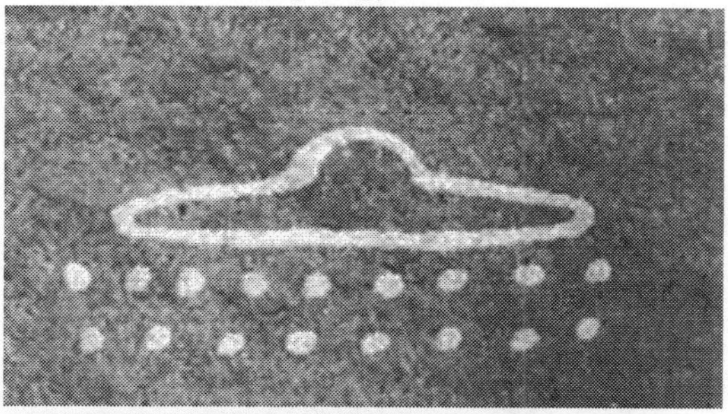

Author Milton Brenner describes engravings from French cave art depicting UFOs 12,000 years ago.) [159]

Fakes and hoaxes

But there is a long history of archeological fakes too. [160] Here are three possibles: There is always a chance they could be real, but I have been unable to corroborate any sufficiently to assure authenticity. Two Egyptian hieroglyphs look remarkably like space vehicles [161] [162]

[158] Pakal the Great, 17 feb, 2022, https://www.chichenitza.com/blog/pakal-the-great

[159] Milton Brenner, "Pale Skin, Giants, and the Great Transition," Trafford Publishing, 2013, see illustration 7.

[160] Jane MacLaren Walsh, Archeologie's hoaxes, fakes and strange sites, Archeology, May 2010. http://archive.archaeology.org/online/features/hoaxes/

[161] "UFO and alien hieroglyphics," https://i.ytimg.com/vi/DXT6BxhC_V8/hqdefault.jpg

[162] Ibid.

There is also an alleged UFO pictograph in the famous French caves of Niaux, (above) but, again, after perusing their photo libraries for hours, I could not find it.[163]

Not convincing.

Examples with a higher level of authenticity

In Mayan culture, there are statues of helmeted creatures who appear to be in either in elaborate headdress or space suits. 164

In Valmonica, Italy there is a 10,000-year-old petroglyph which seems to show helmeted beings.165 Below we see a helmeted being from Guatemala (left) and another from Australia.166

[163] The cave paintings of Niaux France, Bradshaw Foundation, http://www.bradshawfoundation.com/niaux/

[164] Ancient Astronaut, Mayan, https://www.pinterest.com/pin/488148047093806233/

[165] Wikipedia Ancient Astronauts, https://en.wikipedia.org/wiki/Ancient_astronauts

[166] Tumblr https://64.media.tumblr.com/d0e4ebdaf0ec170108c1e6f5e73157c2/tumblr_p4r0kdwekm1ugzxkmo1_1280.jpg

In Australia, 10-16,000-year-old rock art at Kimberly depicts helmeted creatures.

African cave art displays many large-headed figures and creatures which sport antennae or helmets. [167]

A history text on the history of Babylon, written in 250 BC by Borosus, described lizard-like beings. Not far from Babylon is the Ubaidian culture where these lizard-like humanoid figurines were found.[168]

[167] Bruno Classens, African Rock Art, 2013 https://brunoclaessens.com/2013/08/african-rock-art/

[168] Carter, Robert A. and Philip, Graham Beyond the Ubaid: Transformation and Integration in the Late Prehistoric Societies of the Middle East (Studies in Ancient

Reptilian figurines from 5500 BC found near UR and Eridu

Oriental Civilization, Number 63) The Oriental Institute of the University of Chicago (2010). Source of the photos

Egyptian hieroglyph from the wall of Hathor shows an ape-like or lizard-like figure (this time with a tail) in what seems to look like a pedagogical pose.

Utah rock art depicts creature with antennae (right).

The British Museum holds a collection of Sumerian artifacts which strangely look like big-eyed humanoids.

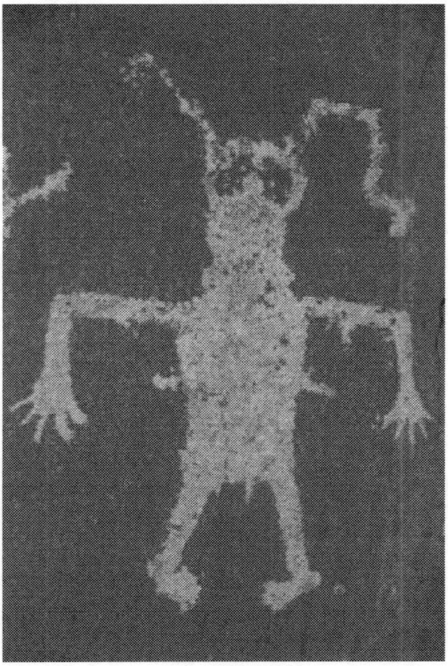

From Predionica, Kosovo. Vinca-Plocnik Culture, Late Mesolithic (5th mill. BCE). Height 16 cm Inv. 158 Belgrade National Museum[169]

Rock art from the McGee Springs, Barrier and Wild Horse Canyon in Utah and Colorado show beings repeating the large-eye theme.[170]

[169] *Chapman, John (1981). The Vinča culture of south-east Europe: Studies in chronology, economy and society (2 vols). BAR International Series. 117. Oxford:*

[170] Rock Art Petroglyphs, http://www.crystalinks.com/petroglyphs1.html

Similar petroglyphs of helmeted humanoids with antennae are found in Niger and thought to be 7000-11,000 years of age. This could be something more prosaic, like women with headdress holding a bag (lower right) but one cannot help but see the similarity in the antennae-like protrusions of the Niger petroglyph with the Utah figures.[171]

[171]The authenticity of the Niger rock art was confirmed for me by Prof Alexander Antonitis of the Dept of Archeology & Anthropology at the University of Pretoria; personal email.

There are debunkers who believe many of the artifacts UFO enthusiasts cite have been photo-shopped or hoaxed, but I have not encountered any genuinely convincing arguments with the specimens cited above. [172] [173] [174]

A search of academic databases on such hoaxes has not revealed any commentary on the particular specimens shown here either.

Below we have more entries:

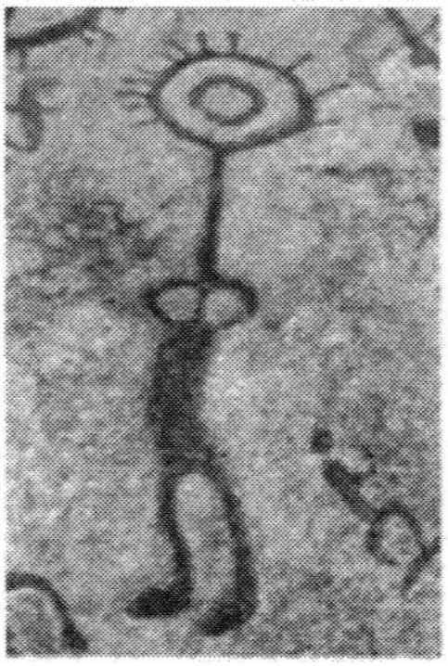

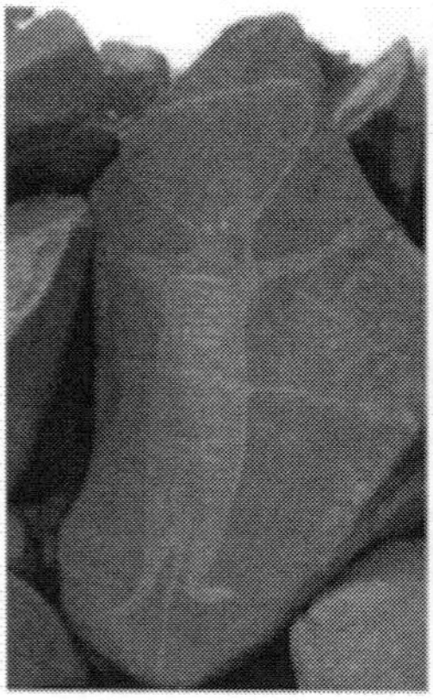
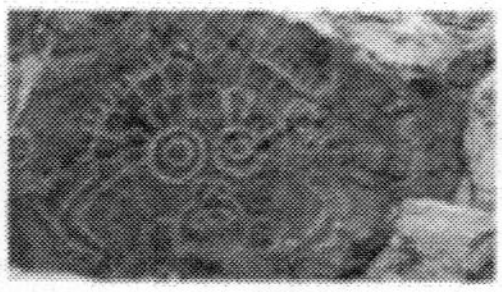

[172] "The Abydos hieroglyph does not depict a helicopter," Ibid. http://members.tripod.com/~a_u_r_a/abydos.html

[173] Archeology Hoaxes,

http://hoaxes.org/archive/display/category/archaeology

[174] James Porter, "The Mayan Hieroglyph hoax," 2013 Estudios De Cultura Maya, 20. https://doi.org/10.19130/iifl.ecm.1999.20.448https://revistas-filologicas.unam.mx/estudios-cultura-maya/index.php/ecm/article/view/448

Upper left, Canada, middle right and upper right Vantage Washington, Ginkgo petrified Forest, lower left, Barrup Rock, Australia, middle Inner Mongolia, lower middle, Arizona, lower right, Tassil Niger (10,000 BC)

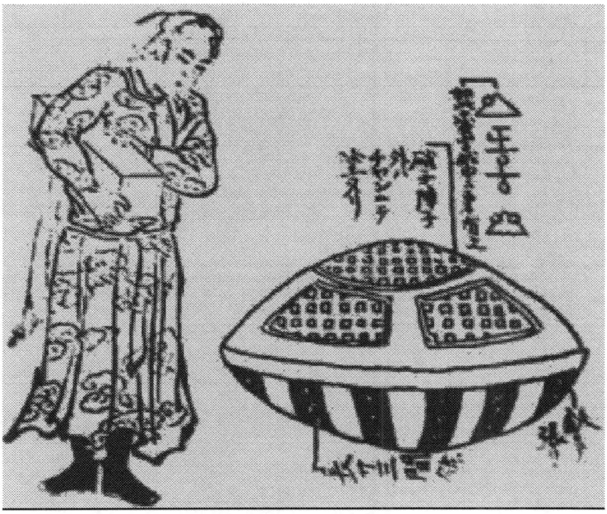

In Japan there is a drawing from the book "Ume No Chiri" (Dust of Apricot) that tells of a foreign ship and crew which was witnessed at Haratonohama, Japan. "The outer shell was made of iron and glass and strange letters were seen inside the ship." [175]

There are also representations in art. A 1583 tapestry known as "A Summer's Triumph" clearly shows a UFO in the sky.[176]

[175] Ancient Code Team, "Utsuro Bune, The Legend of the Hollow Ship," 20 May, 2024, https://www.ancient-code.com/utsuro-bune-the-legend-of-the-hollow-ship/

[176] "The enigma of disc-shaped objects in Summer's Triumph Tapestry,"Iraq Ancient Origins Tours, Public Domain, http://www.ancient-origins.net/unexplained-phenomena/unidentified-flying-objects-sky-summer-s-tapestry-001118

In Kosovo, there is a fresco from the 14th century in Visoki Decani Monastery showing individuals in flying machines. This is certainly rather prima facie evidence, but mainstream art historians prefer that these objects be interpreted as "anthropomorphic symbols of the Sun and the Moon."[177]

Hmmm: beings traveling in space in vehicles, or symbolic representations of the sun and moon?

Are we perhaps witnessing a manifestation of that seemingly incurable academic proclivity toward denial?

Another more modern entry (AD 1495) is from the Fitzwilliam Museum, a painting of The Baptism of Christ showing a strange UFO-like object beaming down on Christ (right).

India cave painting

In 2010, in the Hasangabad district of Madhya Pradesh, India, an unusual cave painting was discovered. It looks very much like a being in some helmeted suit and "a classical flying saucer-shaped UFO that appears to be

[177] Digital Hammurabi, "Ancient Aliens Debunked," 2020 https://www.youtube.com/watch?v=2YYxgLVnr7E

either beaming something down or beaming something up...A force-field or trail of some sort is seen at the rear of the UFO." [178]

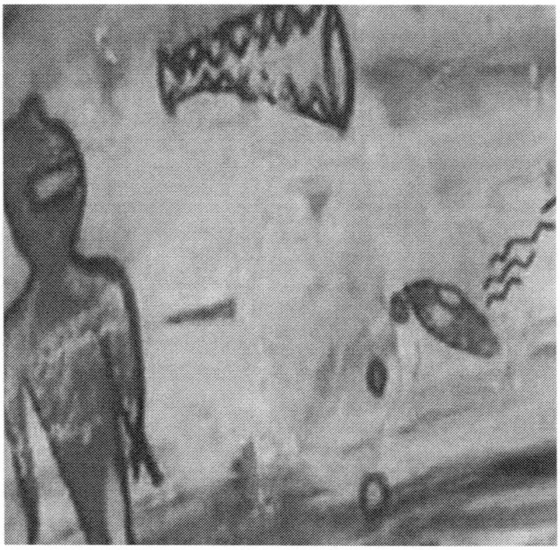

Rock art of space vehicle in Charma, India

It is always possible, of course, that these cave paintings might have been hoaxed in order to increase eco-tourism or archeological reputation, and perhaps one day we will all have to blush over this. This specimen, though, has been authenticated by a local archeologist, Wassim Khan, and there do appear to be citations for a Mohammed Wassim Khan in other publications;[179] he lists himself as affiliated with the "Archeological Environment Research and Tribal Welfare Society," and that similarly appears to be real. Further, I confirmed he was a guest faculty member at Government College.

After four months of searching, I finally found Wassim Khan, corresponded with him, and believe the rock art is not only real, but not of recent

[178] Rashmi Drolia, "10000 year old rock painting," India Times, July 2014, http://www.archaeologydaily.com/news/201002173333/Prehistoric-UFO-and-ET-images-found-in-remote-cave-in-India.html The reader should be cautioned that this discovery is so new, it has not been vetted and for all we know it could be contemporary graffiti or a hoax. As for Wassim Khan's other archeological work, see http://www.rediff.com/news/slide-show/slide-show-1-flying-dinosaur-fossil-found-in-madhya-pradesh/20100608.htm

[179] "Monkey Kingdom," 12 April 2014, bharatuntoldstory,https://www.google.com/search?q=Wassim+KhanArcheological+Environment+Research+and+Tribal+Welfare+Society&ie=utf-8&oe=utf-8

vintage. [180] [181] [182] I was not able to get a date for this rock art out of him, however, and I am still trying, but I do have his website and he appears to be a legitimate archeologist. [183]

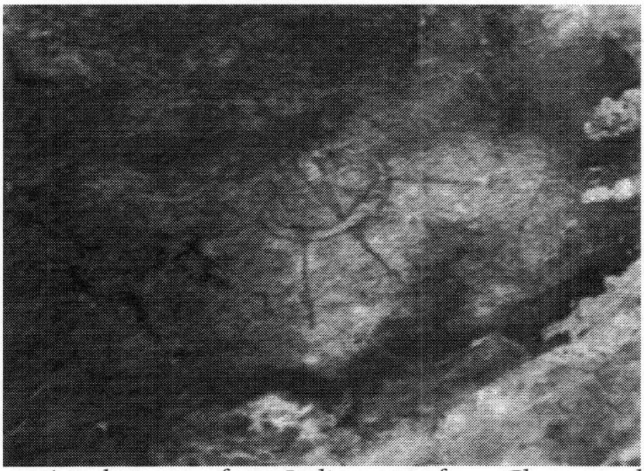

Another entry from India comes from Charama, the Chhattisgarh State Department of Archaeology which reports of a cave under the villages of Chandeli and Gotitola (right).

Archaeologist J.R. Bhagat says these paintings are 10,000 years old and appear to depict a crude drawing of a space vehicle. [184] The legitimacy of these specimens is quite certain.[185]

[180] "UFO sightings, cave paintings," History, http://coolinterestingstuff.com/ufo-and-ancient-astronaut-prehistoric-paintings-found-in-remote-cave

[181] Pratik Chheda , "Vidic café," 01 Mar, 2015, Scribd, https://www.scribd.com/document/257282806/Vedic-Cafe-the-Black-Cave-Wormhole-of-Ramayana

[182] Amina Khan, "That's no bird," 15 July 2014, Los Angeles Times, https://archaeologynewsnetwork.blogspot.com/2010/06/flying-dinosaur-fossils-found-in-madhya.html#bleoQ6sjOsfoQlvM.97

[183] Sascha Pare "13 of the oldest archeological sites," Live Science, 9 Oct. 2023, https://www.ourarchaeology.com/p/about-us.html

[184] Rashmi Drolia, Ibid.

[185] Ibid.

So is the rock art in Africa showing a very similar kind of drawing of a

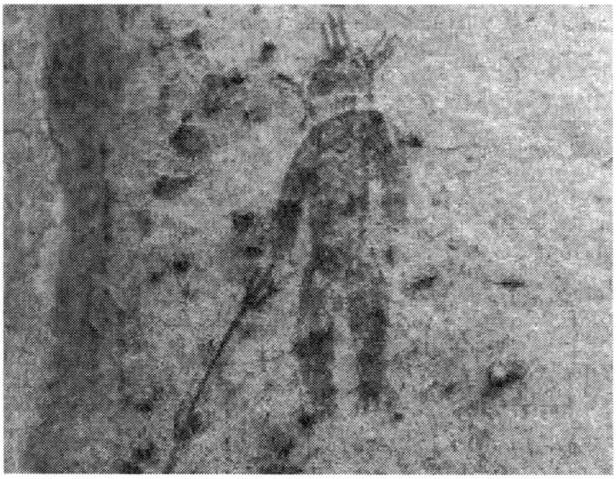

vehicle.[186] Another from the same collection in India seems to show a helmeted being.[187]

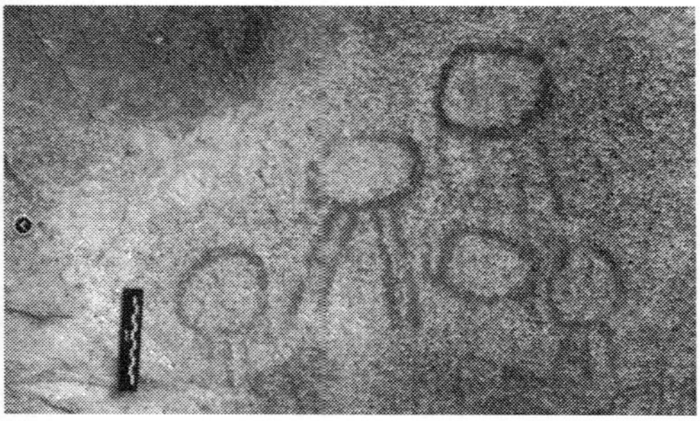

Moab Utah, creature with 4 fingers. Source: https://www.climb-utah.com/Moab/owl.htm; https://de.123rf.com/photo_17726414_buschm%C3%A4nner-san-felsmalereien-von-antilopen-drakensberge-s%C3%BCdafrika.html

Despite possibilities of hoaxes, there is a thicket of authentic archeological specimens worthy of serious attention: the Nazca lines showing a 32 meter humanoid visible only from the sky, hieroglyphs, figurines, ceramics,

[186] Trust for African Rock Art http://africanrockart.org/

[187] Jason Colovitto, "Times of India endorses ancient aliens theory," July 16, 2014, blog, https://www.jasoncolavito.com/blog/times-of-india-endorses-ancient-astronaut-theory-says-cave-paintings-show-aliens-and-ufos

rock art and medieval art of alien-like creatures, lizard-like beings walking, helmeted creatures with an tense, alien-like figures with four fingers, large-eyed creatures, large headed bipedal beings, and multiple drawings of extra-terrestrial craft or paintings of flying beings that date from as early as 13,000 BC well into the Mayan era (950-1539 AD)—spanning, in other words close to 15,000 years.

The most basic question is clear: Is it possible our remote ancestors were trying to tell us they had visitors?

Chinese, Mexican & Egyptian Pyramids

There are pyramids in China which most Americans never heard of. In 1945, a US Air Force pilot flew over them and took pictures.

Chinese pyramid above and Teotichuahan pyramid below

The Chinese pyramid, for example, is said to be more 5,000 years old—more reasonable voices say 2000 years of age. [188] [189] It is the tallest pyramid in the world at over a thousand feet.[190] Its design is similar to the Pyramid of the Sun in Mexico, estimated to be 2,000 years old. Teotihuacan's Pyramid of the Sun is then the third tallest pyramid in the world, made from over three

[188] "The White Pyramid of Xian," Cyberspace Orbit, No date given, http://www.bibliotecapleyades.net/piramides/esp_piramides_china_3.htm

[189] "China Great Pyramids Controversy, "Eye of the Psychic, 2024, http://www.philipcoppens.com/china_pyr.html

[190] Walter Hain, Great White Pyramid Discovered, Mahain website, October 2006, http://www.bibliotecapleyades.net/piramides/esp_piramides_china_5.htm

million tons of matrial.[191] The structure was then layered with stone and then coated in brilliant-colored plaster. The plaster is now gone. The pyramid encrypts pi in its construction (more on that, below). It was built allegedly by the Teotihuacanos, whose hieroglyphic language has yet to be decoded.

'Despite our having timidly ventured into the waters of 'forbidden archeology, think about these connections:

How unreasonable would it be to suggest that similarities of these unusual structures might demand a more global theory with regard to their origin? Mainstream archeology prefers to see them as entirely indigenous to the peoples and who created them and as separate and discreet entities.

But we can't help wondering how the pyramids of Egypt, Mexico, and China might be related to one another.

Mythology certainly offers a possibility. As the Egyptians traced their origins to the stars and Osiris, Chinese lore explains the construction of its pyramids coming from the age of the "old emperors" who reigned in China. These ancients did not originate on earth. They were the "sons of heaven who came down to this planet on their fiery metallic dragons." [192] Just as the Chinese reference the "sons of heaven," the Egyptians footnote their gods, Osiris and Horus, while the Teotihuacan is associated with the gods *Quetzalcoatl, Xipe, Totec* [193] ... all star-deities mysteriously connected to these monuments.

Giza pyramid plane in Egypt

Robert Bauval, an American engineer working in Saudi Arabia, often took his children to Egypt. A surveyor by trade and deeply fascinated by the pyramids, he enjoyed camping near the base, but found the layout of the three pyramids near Giza illogical. They were built to incredibly exacting standards but *mysteriously off-center*. He could not understand how Egyptian

[191] Temple of the Sun. By: Pérez de Lara, Jorge, Archaeology, 00038113, Nov/Dec2005, Vol. 58, Issue 6

[192] Hartwig Hausdorf, China Alpha Film Series, www.alphaomega.se/english/chinesepry.html.

[193] Mark Cartwright, "Teotihuacan," 20 Feb, 2015, World History Encyclopedia http://www.ancient.eu/Teotihuacan/

engineering could be so precise and yet the locations of the pyramids *skewed*... unless, of course, they were skewed intentionally.

The Belt of Orion

One evening, he was lying in the sand and looking up; he had a mind-boggling insight. He could see that the constellation Orion, directly above, was off-center too and did not fit a straight line. He measured the displacement and skewness and then spent the larger part of his life examining the analogous similarities in the organization plan of the three pyramids. His findings were so accurate and intriguing that Egyptologists published his findings even though he had few credentials beyond a civil engineering background to gain access to these elite journals.[194]

[194] R. Bauval and A. Gilbert The Orion Mystery. New York: Crown, 1994.

Photo of Giza pyramids appearing in this section, courtesy of the Institute of Collective Behavior and Memory.

The location of the three major pyramids in Giza are "off" by an amount which parallels the skewness of three stars in Orion. The pyramids, he concludes, are like dropping a plumb line down from the stars, as if they were an attempt to make a heavenly city on earth. The theory goes further.

What were originally thought to be airshafts within the pyramid focus themselves exactly on Sirius on two dates, he says, every 13,000 years separated from the next. If this extraordinary theory is correct, and the shafts were set to align with Sirius on those dates, *then* the Egyptians most certainly knew about precession (the wobble effect) of the planet and that it would take roughly 13,000 years before the next alignment with Sirius was setgles between Orion and the Giza pyramid plane by mathematicians Vincenzo Orofino and Fisica de Girodi of the University del Salento in Italy. Their summary corroborates Bauval and is given below:

> The results found in the previous sections can be summarized as follows: a) the relative positions of the three Giza pyramids coincide, within the uncertainties of the naked eye astrometric measurements, with the relative positions of the three stars of the Orion Belt; b) in the scale that would have been chosen by the pyramid builders to reproduce the Orion Belt on the ground, the linear distance between the Menkhaure pyramid and the Nile, along the straight line connecting the two extreme pyramids of Giza, practically matches the angular distance between Mintaka and the central point of the Milky Way along the straight line connecting the two extreme stars of the asterism; c) the visual magnitude of the stars of the Belt is presently correlated with the height of the corresponding pyramids evaluated a common reference level (i.e. the base level of the Khufu pyramid). Since the star evolution models suggest that the magnitudes of all the three objects of the Belt at the time of the pyramids were substantially equal to the present ones, the above- found correlation was still valid at that epoch.[195]

We started this discussion, however, thinking that the pyramids across the Earth may have something in common with each other. What is absolutely intriguing is that the pyramid layout in Egypt *is very similar to the skewness of the pyramids in China,* as if both sets of structures parallel the dogleg turn of the Belt of Orion—and that is where things start feeling incredibly weird. (See below)

[195]Vincent Orofino, "A quantitative astronomical analysis of the Orion Correlation Theory, https://arxiv.org/pdf/1109.6266.pdf

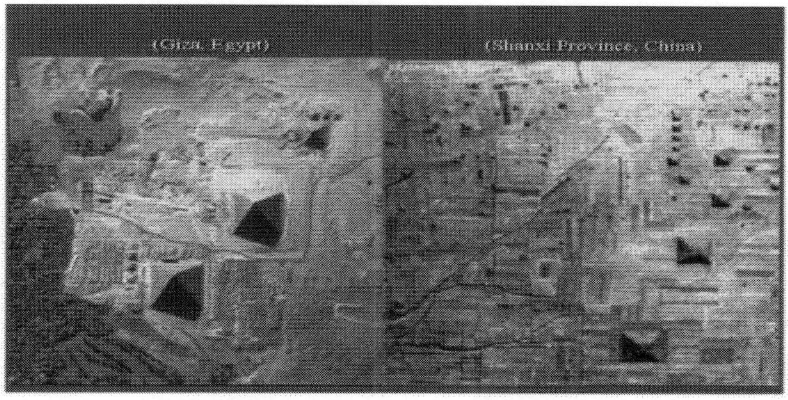

Pyramid planes of Giza (left) and China (rt)

A far more precise study was made of this apparent correlation in anIf one looks at the Chinese pyramids and draws a line through the largest, next largest, etc., the smallest, also appears to form a 15-degree angle which reaches out dogleg just like the Giza pyramids and the belt of Orion.

It gets worse!

Since these initial observations were made, enterprising—and open minded—researchers noticed that the pyramid plot plan in Mexico could also be oriented to the Orion Belt. Here, then, is the third pyramid system to join in this increasingly loud chorus (see below) [196]

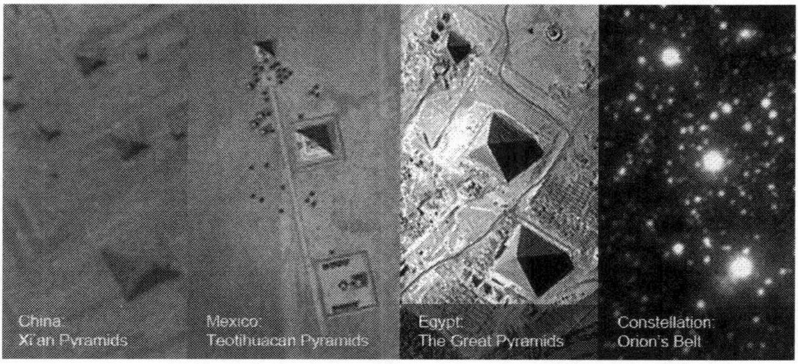

Aerial view of pyramid plain in Egypt (Left), in China (middle), and Belt of Orion (far right)

[196] "Ancient Code," http://www.ancient-code.com/orion-constellation/ See also: Robin Heyworth, Teotichuan: pyramid of the sun mystery, 11 Nov 2014, Uncovered History, http://uncoveredhistory.com/mexico/teotihuacan/teotihuacan-pyramid-of-the-sun/

This is where it becomes positively mind-blowing. Separate cultures of antiquity, thousands of miles from each other, and thousands of years from each other, built some of the largest and most unusual structures on earth, all ostensibly plumb-line analogs to the Orion Belt and all in countries with no apparent connections or communications between each other

Regarding the uncanny alignment of the Teotihuacan pyramids,

> The Pyramid of the Sun's size is such, that it immediately draws parallels to the Great Pyramid of Khufu in Egypt. In fact, almost unbelievably, the base of the Pyramid of the Sun is virtually identical in size to that of the Great Pyramid of Giza (which measures 230m). And it isn't just the size of the Pyramid of the Sun that strikes accord, for it also sits amidst a trio pyramids that appear identically aligned to those of the Giza Plateau. At Teotihuacan, these pyramids are known as the Pyramid of the Moon (furthest north), the Pyramid of the Sun (central) and the Pyramid of Quetzalcoatl. Furthermore, the alignment at both sites clearly reflects the alignment of the three stars of Orion's Belt.[197]

Now, if this isn't enough, another bombshell drops upon us from the pens of researchers Christopher Knight and Alan Butler who report their findings in *Before the Pyramids.* They found yet another such monument, this time in Neolithic England.

A "henge" is an earthen mound or an earthen structure. One ancient henge was found in Thornborough and it traces out another analog to the Orion Belt. Butler used photos of Orion's Belt, blew them up, put them into his computer program, and compared the Google Earth image of Thornborough with Orion:

> I drew lines as I had done with the stars and then increased the lines proportionally until the longer line...was 366 cm long on the drawing program. I then noted that the shorter line. ...was exactly (not nearly, not very nearly, but quite exactly) a tiny bit under 360 cm. Result. The Thornborough henges are not a good copy of Orion's Belt, they are not even a very good copy of Orion's Belt. They are an exact, absolutely, absolutely, absolutely *exact copy* of Orion's Belt. [198]

At this point my head was spinning, so I wrote to a non-ancient-alien oriented archeologist for some guidance and perhaps a different and more sobering point of view. He promptly reprimanded me, a psychologist, for even being interested in a subject outside my wheelhouse. Although he only

[197] Heyworth, Ibid.,

[198] Christopher Knight & Alan Butler. Before the Pyramids. London: Watkins Publishing, 2009, p. 69

had a BA, he had impressive publications on the Thornborough Henge and accused the authors of *Beyond the Pyramids* of "popularization."

So, somewhat chastised and thus a bit more suspicious of a popularized text, I thought I would buy a compass and check out these, perhaps, overly enthusiastic authors. I measured the angles and indeed it comes out to rather exactly 15 degrees of arc. Surely these are amateur observations, but they are consistent with the more serious work of mathematicians Vincenzo Orofino and Fisica de Girodi cited earlier.

So, I wrote to Dr. Orofino a second time, this time about Thornborough, and he said that he found the middle star (Alnilam) a bit too close to the center, but otherwise "this not a conclusive argument against a putative correspondence between these structures and the asterism of the Orion Belt."[199]

In this investigative storm, two more academic peer-reviewed papers appeared, and both suggested Thornborough was related to the Orion Belt.[200] In other words, it is not a preposterous suggestion at all.

[199] personal email.

[200] Jan Hrding, Ben Johnston, & Glyn Goodrick, "Neolithic cosmology and the monument complex of Thornborough, North Yorkshire," Archeoastronomy, University of Texas Press 2006. https://www.academia.edu/1376624/Archaeoastronomy?auto=download

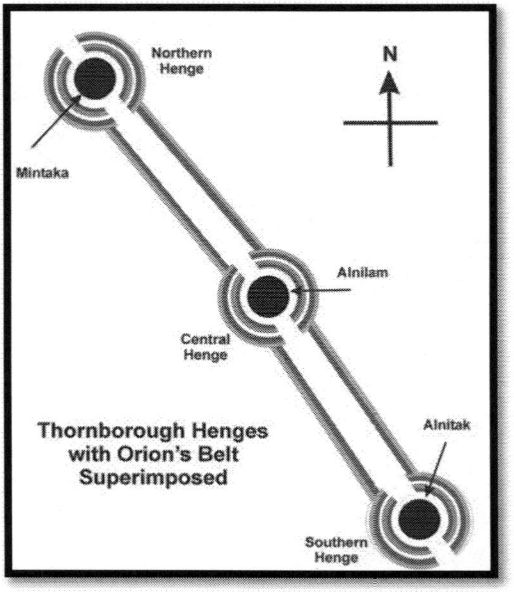

Here is a more systematic work that shows the rather exacting portrait of Thornborough paralleling Orion. [201]

We don't know what particular sky-gods Thornborough was connected to, but a nearby lead bust of the Goddess Isis was found in Wiltshire only 55 miles away. [202] Isis was the sister-wife of Osiris in mythology, so we are not being unduly speculative in thinking that Thornborough might have been related to Osiris as well as Orion. The thought here is that the Chinese Pyramid plane, the Giza pyramid, the pyramid plot plan of Teotihuacan as well as the Thornborough Henge *all seem related to the Belt of Orion with their mutually shared angular doglegs.*

Some estimate the age of the Thornborough Henge was 2,500-3,000 BC. [203] That is roughly the period when the Great Pyramid of Giza was

[201] Sigourd Towrey, "On Earth as it is in Heaven," Frontiers in Archeology and History, no date, http://www.ancient-origins.net/ancient-places-europe/sacred-prehistoric-neolithic-complex-thornborough-henges-003652

[202] Jorn Schuster, "A lead bust of the goddess Isis,"

04. May, 2011 Cambridge Core, https://www.cambridge.org/core/journals/britannia/article/a-lead-bust-of-the-goddess-isis-from-groundwell-ridge-swindon-wiltshire/354264AF1E0A318193A7FD17599CE61A

[203] Prehistoric Yorkshire, The Yorkshire Archeological Society, no date, http://www.prehistory.yas.org.uk/content/thornborough.html

constructed. However, another study dates the Thornborough Henge 1,000 years *earlier than* the Great Pyramid! [204]

Is it merely a coincidence that disparate cultures, each unaware of the other, built similar structures, in China, Mexico, and in Egypt—and now in England—following similar plot plans?

All of this, to believe conventional archeology, happened entirely by coincidence and without any contact between cultures. Moreover, the fact that these engineering feats are attributed to the heavens or the influence of "the Gods" is abjectly ignored in academia. It is disdainfully relegated to what is known as "pseudo-archeology." Seems at the very least curious!

The author standing in front of the Pyramid of the Sun in Teotihuacan, Mexico.

- The pyramids in Egypt, China, and Mexico (Pyramid of the Sun) are similar in design.
- These pyramids are some of the largest structures made by humankind at the time.
- All three of these pyramid locations are associated with sky gods.

[204] "Henges older than Pyramids," BBC Feb, 2006, News,http://news.bbc.co.uk/2/hi/uk_news/england/north_yorkshire/4713820.stm

- The pyramid planes (Giza, Mexico, and China) appear to reference the Orion Belt.
- The Thornborough Henge may be associated with heavenly deities too, implicating Osiris, and also appears to be an analog of the Orion Belt.

Those concordances and connections do not appear to impress the mainstream at all. It is all thrown into the dumpster of "speculation," and none of these correlations are at all relevant or worthy of further study.

It makes one wonder if perhaps somewhere there is a God of Denial where academic acolytes seeking tenure—who might have been interested in these connections—are regularly sacrificed.

Osiris, Orion, and Dolichocephaloids

It is certainly legitimate to wonder what all these monuments have to do with the Orion Belt. If the Thornborough Henge points to Orion, as well as all these pyramids, who could have come from there? The stars in the Orion Belt are between 243 to 1350 light-years from Earth, so if our stellar visitors could travel at close to the speed of light, it would still take close to *half a millennium* to make a round trip from the nearest star. [205]

Did they find a time-warp wormhole to traverse?

But before capitulating to mainstream archeology's insistence that it is all fable and nonsense, let us pay a short visit to the Papyrus of Turin. It is displayed at the museum of Egyptology in Italy, and the manuscript has been dated to roughly 1150 BC. [206]

What is unique in the document is the list of Kings. It purports to list virtually all the kings and pharaohs of Egypt like Meni, Teti, Qenti, Hepu, Seth. etc., roughly 255 names covering 17 dynasties.[207]

One would think this might cover a millennium or two, but curiously the list of kings begins with *heavenly deities* who first founded and ruled over Egypt. It then lists "demi-gods," and finally the more secular 255 pharaohs.

The founding fathers—sorry gods—are quite numerous and include Amun, (Amen), Anubis, Horus, Osiris, Nut, and others. Quite astonishingly, the reign of all the gods, demi-gods and pharaohs spans a time from the

---

[205] Mary Lou Whitehorn, "How far are the constellations," Space, 24 Jan. 2007, http://www.space.com/3380-constellations.html

[206] "The Turin Erotic Papyrus," Open Culture. 9 July 2014, http://www.openculture.com/2014/07/the-turin-erotic-papyrus-the-oldest-known-depiction-of-sex-circa-1150-b-c-e.html

[207] Wikipedia, The Turin King List, https://en.wikipedia.org/wiki/Turin_King_List

beginning of Egypt until the time the papyrus was written, and that number is a staggering 13,420 years!

Say what?

> The ancient Egyptians viewed their civilization as a legacy coming directly from divine beings who existed in Egypt thousands of years before the pharaonic dynasties that we know about. The papyrus of Turin (or more exactly the royal canon displayed at the museum of Egyptology of Turin), written in hieroglyphics and dating back to Ramses II, presents a list of all the pharaohs who reigned in the land of Egypt. This list includes not only the historical pharaohs, but also the "divine pharaohs who came from elsewhere" who reigned before the first dynasty of Menes. We are also told that this preceding lineage reigned for 13,420 years![208]

Most important to our interest is that one of those gods, Osiris, is clearly equated with Orion. In the Pyramid Texts written approximately in 2,400 BC (about the time of the construction of the Great Pyramid) it says: "Behold him who is come as Orion! Behold Osiris who is come as Orion" [209] It goes on to say,

> Hail O King! Thou art this great star, the bearer of Orion, who ferries the ship of heaven with Orion...This King has come. This King glories Orion. This King exalts Osiris. This King places the gods on their thrones.[210]

Certainly at least one Egyptian God, Osiris, is thematically connected with Orion, so perhaps it is the god, Osiris, who is the reason these structures are earthly analogs of the Orion Belt.

But what about China?

Surely, if Osiris was not a fantasy, but a real star-being who was part of the pantheon of gods that founded Egypt many millennia before we ever thought they did, how do we explain the Xian pyramids which also seem to point to the Orion Belt in far away China?

Strangely, in 2016 a geochemist, Sun Weidong, in China argued that the very beginnings of China were not started by the Chinese at all, but by Egyptians!

> In the past year, Sun, a highly decorated scientist, has ignited a passionate online debate with claims that the founders of Chinese civilization were not in any sense Chinese but actually migrants from Egypt. He conceived of this connection in the 1990s while performing radiometric dating of ancient Chinese bronzes; to his surprise, their

[208] "Egypt before the Pharoahs," Gigal Research , 2013, http://www.gigalresearch.com/uk/publications-pharaohs.php

[209] cited in John Gwyn Griffiths, The Origins of Osiris and his Cult, Brill 1980

[210] Ibid.

chemical composition more closely resembled those of ancient Egyptian bronzes than native Chinese ores....According to him, its bearers were the Hyksos, the Western Asian people who ruled parts of northern Egypt as foreigners between the 17th and 16th centuries B.C... He notes that the Hyksos possessed at an earlier date almost all the same remarkable technology — bronze metallurgy, chariots, literacy, domesticated plants and animals — that archaeologists discovered at the ancient city of Yin, the capital of China's second dynasty, the Shang, between 1300 and 1046 B.C....Of the 200 or so items of bronze ware he was responsible for analyzing, some came from the city of Yin. He found that the radioactivity of these Yin-Shang bronzes had almost exactly the same characteristics as that of ancient Egyptian bronzes, suggesting that their ores all came from the same source: African mines.[211]

This is certainly taking us on one wild ride of conjecture, but these are the only attempts that I have found which might offer to explain or link why we find gigantic monuments across the globe which appear to be pointing the Belt of Orion and which date to the very dawn of human culture and civilization.

[211] Ricardo Lewis, "A decorated scientist has ignited a passionate debate with claims that the founders of Chinese civilization were not Chinese," Business Insider 06 Oct 2016, , cited in http://www.businessinsider.com/a-decorated-scientist-has-ignited-a-passionate-debate-with-claims-that-the-founders-of-chinese-civilization-were-not-chinese-2016-10?utm_source=feedburner&utm_medium=referral

Dolichocephaloids

The God Horus is depicted in Egyptian hieroglyphs as a bird or falcon who is bipedal and sometimes showing an elongated skull(rt.) [212] Osiris too is portrayed as having an elongated skull or a conical headdress.

There is a compelling history of elongated skulls and cone-headed figurines in mythology, as we have already seen, but there are *actual* elongated skulls, which have been found over the whole earth from Peru to Siberia, and dating to early in human history.

Mythology meets reality!

The condition of an elongated skull is technically referred to as a dolichcephaloid, but this is not a physical aberration, but a cultural practice of boarding a child's skulls:

> "It is therefore possible, by lashing a rope around the head, with a board placed at the back of the skull, and perhaps the front as well, to alter the shape of the head over time. Many authors state that the time period to perform this shaping was about 6 months to 3 years, but since the practice is no longer performed to my knowledge, no one really knows. Examples of this technique, supposedly last performed on infants in the Congo of Africa and the Island of Vanuatu in the south Pacific Ocean area known as Melanesia, well into the 20th century, have also been found in Egypt, during the Amarna period, Sudan, Iraq,

[212219] Anne Baker, Oude Egyptenaren Schouwden de kiomst van Christus, 23 06 2018, Sekem, http://www.sekemvrienden.nl/oude-egyptenaren-schouwden-komst-christus

Syria, Russia, the island of Malta, as well as many places in Peru and Bolivia, and amongst the Olmecs of Mexico." [213]

Numerous Egyptian royalty mummies have been unearthed displaying elongated skulls.

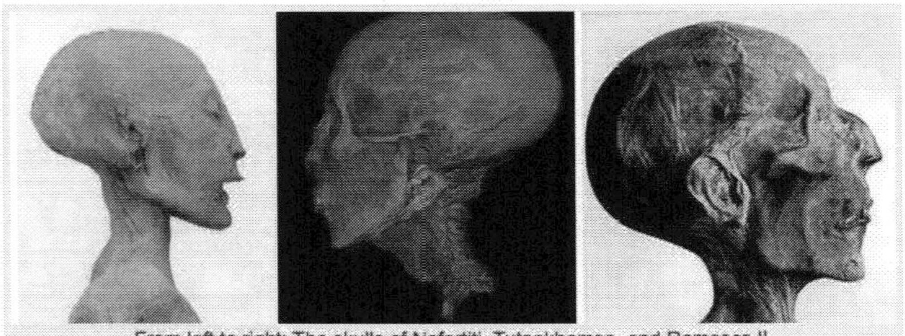

From left to right: The skulls of Nefertiti, Tutankhamen, and Rameses II.

Below, one can see the same sort of elongated skull unearthed in Siberia. And yet others in Peru. [214]

[213] "Elongated skulls of Peru",9 Apr, 2013 Ancient Origins https://www.ancient-origins.net/ancient-places-americas-opinion-guest-authors/elongated-human-skulls-peru-possible-evidence-lost

[214] "Paracas Elongated Skulls of Peru, No author or date http://www.latinamericanstudies.org/paracas-skulls.htm

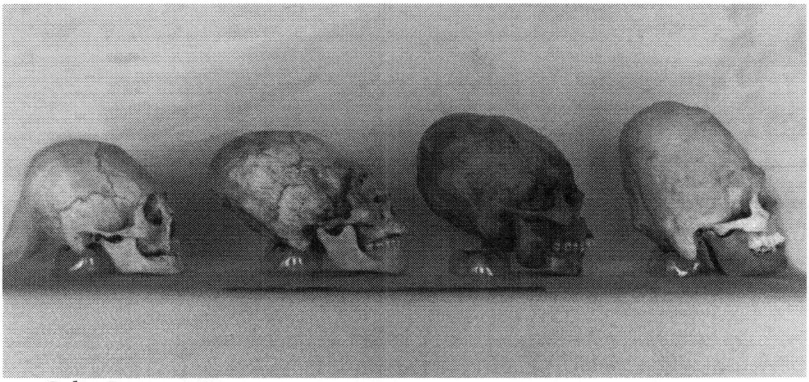

Colin Baras writes,

> They are global, found in many cultures, thousands of kilometers apart...The occurrence of elongated heads can be found over long periods of time throughout history...Even more interesting they most often occurred in the elite: royalty, high priests/ priestesses, and nobility in their respective cultures...They have also been recorded in pre-historic *Homo sapiens* (12,000 BCE), from the Shanidar Cave in Iraq. . .in Neolithic Southwest Asia. We see them in the Old World, Huns and Alans, and Eastern Germanic tribes. Hippocrates describes macrocephali, or elongated heads, in 400 BCE. Indigenous peoples of Australia, Maya, and North American Indian tribes (Chinookan of the Northwest and Choctaw of the Southwest) also had elongated heads.[215] [216]

Our first thought, of course, is "are these aliens?" One Van Daniken acolyte, who goes by the name of Brian Forester, says he has attempted to sequence the DNA of these aliens:

> But a few fragments I was able to sequence from this sample 3A indicate that if these mutations will hold we are dealing with a new human-like creature, very distant from Homo sapiens, Neanderthals and Denisovans. I am not sure it will even fit into the known evolutionary tree. The question is if they were so different, they could not interbreed with humans.[217]

[215] Wikipedia, Artificial Cranial Deformation, https://en.wikipedia.org/wiki/Artificial_cranial_deformation

[216] Hippocrates of Cos (1923) [ca. 400 BC] Airs, Waters, and Places, Part 14, e.g., Loeb Classic Library Vol. 147, pp. 110-111 (W. H. S. Jones, transl., DOI: 10.4159/DLCL.

[217] DNA of Paracas elongated skulls of Peru, August 2016, https://www.youtube.com/watch?v=iOfTyc6qb4A

Before gasping with excitement, note that Forester has, from the best I can tell, a bachelor's degree, and his DNA work has not been corroborated from anything I've examined. Others have criticized his claims despite the fact that the UFO ancient alien community is quite atwitter over them. [218]

Real scholarly work which has looked at the Egyptian mummies with elongated heads, has no conclusions yet about DNA, and there is considerable controversy about making any DNA assessments because of contamination issues.[219] [220] *The most definitive study appears to suggest nothing 'alien' in the DNA:*

> Our results, in conjunction with archaeological data, provide substantial evidence that the mummy found in KV55 is indeed Akhenaten and that the KV35 Elder Lady is Tutankhamun's paternal grandmother, Queen Tiye. Moreover, the KV55 mummy and the KV35 Younger Lady mummy can safely be identified as the father and mother of Tutankhamun . . The established Y-chromosomal profiles show identical patterns in Amenhotep III, KV55, and Tutankhamun. This provides evidence that these individuals share the same paternal lineage. Control mummies examined along with Tutankhamun's putative family members yielded different Y-specific alleles...By evaluating the segregation of alleles through the generations, we reconstructed the most plausible royal pedigree: a five-generation family tree.[221]

In examining Tutankhamun's skull the investigators tried to learn if he and his family suffered from "Marfan Syndrome" or abnormally enlarged heads. This might be a biological condition rather than a cultural practice of enlarging the cranium. The investigators did not confirm any diagnosis of Marfan syndrome. [222]

---

[218] Nayell Lomeli, "Fact check: Elongated skulls in Peru are human, not alien," USA Today 19 May, 2021 https://www.usatoday.com/story/news/factcheck/2021/05/19/fact-check-elongated-skulls-peru-human-not-alien-expert-says/4995026001/

[219] Jo Marchant, Ancient DNA, Curse of the Pharaohs, 27 Nature. Apr, 2011, http://www.nature.com/news/2011/110427/full/472404a.html

[220] Richard Wilkinson, The Oxford Handbook of the Valley of the Kings, Oxford University Press 2015.

[221] "Sickness, Hunger, War and Religion, Michaela Harbeck (Ed.), Rachel Carson Center Perspectives, Mar, 2012. http://www.environmentandsociety.org/sites/default/files/1203_sickness_web_color.pdf#page=11

[222] "In the process, however, they discovered a dagger buried with the King which came from a meteorite. It had a gold and crystal handle and the analysis of the ironwork "strongly suggests an extraterrestrial origin."The original article was published in Meteoritics and Planetary Science. Source: Declan Walsh, "King Tut's dagger made of Iron from the sky," New York Times, June 2, 2016.

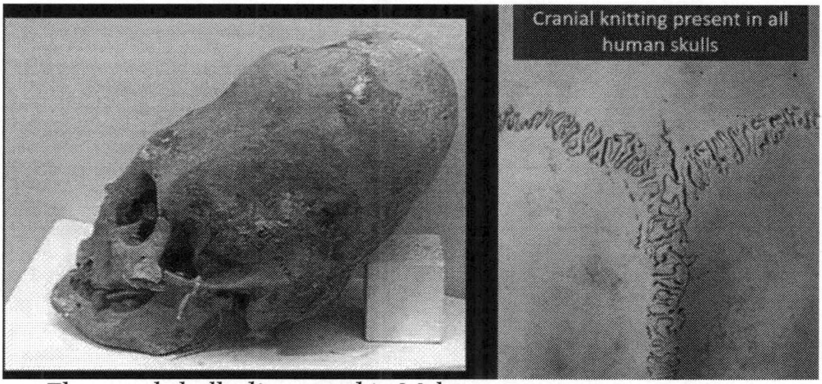

Elongated skulls discovered in Malta

Source: https://www.paulwagner.com/maltas-secret-skulls-the-remnants-of-prehistoric-aliens/

In other words, the elongated skulls were more likely the result of a cultural practice among the pharaonic elite, and the consensus seems to hold that *elongated skulls represented a sign of royalty, divinity, or a connection with a deity. The strange cultural practice date as far back as 13,000 years ago.*

A trip to Malta

As part of my research for this book, I made a trip to Malta. I gave a talk at the university psychology department, but was mainly interested in the Hypogeum, an underground necropolis where over 7,000 remains were found. The architectural wonders there are intriguing because they are older than the pyramids.

I saw rut marks in stone on the island which appear to a naïve observer like someone was carrying goods and supplies up the mountain from the shore or vice versa more than 5,000 years ago. My dismay was immense, however, when they told me the Hypogeum was closed and that I would not be able to see the mystery of the elongated skulls.

When the Hypogeum was first discovered by accident in 1902 the public was not able to see inside until 1985, and then access to the skulls was closed off. Even sporting a Ph.D., I couldn't get to see them.

What is unusual is that these skulls do not show the usual pattern of "cranial knitting" which comes from attempting to form a child's developing skull in an elongated manner. Some have suggested these beings were born this way, and from there the idea that these could be alien rulers began.

We do not have any definitive science that I know of at this writing that would support DNA studies one way or another.

A different approach to interpretation: John Frum and psychoanalysis

What we learn from the John Frum 'cargo cult' of the Caroline Islands is that when a primitive people encounter a being who arrives from the sky and possesses superior technology, there is a tendency to deify the intruder or visitor and construct a religious mythology about him or her, about their return, and about their heavenly ways.

Is it possible that the cultural tradition of shaping an infant's head to have this conical form was a form of *mimicry*, an attempt to fuse with and possess or inherit some of the Godhead's celestial attributes? Social psychologists have long known about mimicry that "people who are less powerful are more likely to copy the movements of those who are more powerful."[223]

Curiously, there are Egyptian texts which "show the expectation that the dead pharaoh would 'become Osiris.'" [224]

Return to psychoanalyst Karl Abraham's notion that in our myths we memorialize important flashpoints in our history with tales or songs that encrypt archaic memories, like "Ring around the Rosy" echoing the Black Death of the 14th century. Perhaps the craniums of early Egyptian pharaohs from Akhenaton to Ramses to Nefertiti were "shaped" as were the craniums of their children designed to similarly memorialize, encode, or reflect their connection, or lineage, to the ancient god Osiris—and thus share in his divinity, to wit, another expression of John Frum mimicry.

Today, we may even see this encoding in contemporary figures. Who might we ask, is most closely associated with the heavens, and the divine, but the pope, the "vicar of Christ on earth?" And when we look at his headdress, as well those of his cardinals, they have an uncanny resemblance to the headdress of Akhenaton four millennia earlier.

[223] Brian Parkinson, Agnete Fischer, et. al., Emotion in Social Relations: Cultural, Group, and Interpersonal Processes. Psychology Press, 2004, p. 141

[224] "The underworld and the afterlife" in ancient Egypt, Australian Museum, Jan., 2023

https://australian.museum/learn/cultures/international-collection/ancient-egyptian/the-underworld-and-the-afterlife-in-ancient-egypt/

Papal mitre (left), Catholic cardinals, and Akhenaten 1336 BC

And Akhenaton, in turn, draws us further back recapitulating the "look" of Osiris as an epiphenomenon in his own elongated skull or skull cap.

The papal headdress called the papal Mitre, dates to Byzantium, but one wonders if students of these artifacts realize it may unconsciously echo far earlier times than that. [225]

Christ and Osiris

Scholars have compared Christ to Osiris. Osiris had over 200 divine names that sound very Christian ("Lord of Lords," "King of Kings," "The Resurrection and the Life," His coming was announced by three wise men which some argue represents the three stars of the Orion Belt.

[225] *Herbermann, Charles, ed. (1913). "Mitre". Catholic Encyclopedia. New York: Robert Appleton Company.*

The Lord's prayer was "prefigured by an Egyptian hymn to Osiris-Amen." Amun/ Amen is the name of an Egyptian god, and it is recited at the end of most every Christian prayer today.

Osiris' resurrection provided hope to all mankind as did Christ's. There are parallels too between Christ and Horus. Both were born of a virgin; both had 12 disciples; both walked on water, etc. [226]

Osiris and Christ were both crucified, both were betrayed, both were the progeny of star people, both rose from the dead. [227] One of the more intriguing parallels between Christ and Osiris is that the birth of Osiris is accompanied by Three Wise Men:

> Osiris' coming was announced by Three Wise Men: the three stars Mintaka, Anilam, and Alnitak in the belt of Orion, which point directly to Osiris's star in the east, Sirius (Sothis), significator of his birth.[228]

And for Christ, we have another three wise men, the Magi, following the Star of Bethlehem, who some believe was Sirius as well. [229] Others disagree and say the Star of Bethlehem was not Sirius but rather the planets Venus or Jupiter.[230]

For those of us astronomically challenged, if one draws a line through the Orion Belt, it points to Sirius, the brightest star in the sky.

Continuing with the Christ-Osiris parallels:

> Astronomer Richard H. Allen states that in European folklore the three stars in Orion's belt are often called the Magi or the Three Kings. And the Christian mythologist, Alvin Boyd Kuhn wrote: 'There is the legend of the 'Three Kings of the Orient' who came on Christmas to adorn the new-born God...The Three Kings were the three conspicuous stars in the belt of Orion that so easily distinguishes this notable constellation and their title was from days of old the Three Kings of Orion. They point almost in a direct line to the following star Sirius, which being the brightest star in the sky was seen as an archetype of the Christ-soul in mankind."[231]

[226] See also Kent Sayler, Journey into Truth, Lulu.com, 2010

[227] D. M. Murdock, Christ in Egypt: the Horus-Jesus Connectin, Stellar Publishing, 2008

[228] D. M. Murdock, "The star in the East and the three kings," Stellar House Publishing 20 Jan, 2020. http://www.stellarhousepublishing.com/star-east-three-kings.html

[229] Ibid

[230] Kim Lawton, "Christmas star debate gets its due on Epiphany." USA Today, Jan 5, 2008.

[231] Is the Jesus Story an astrological allegory, Archeologica, Jan 9, 2008.

Dogon engraving with elongated skulls

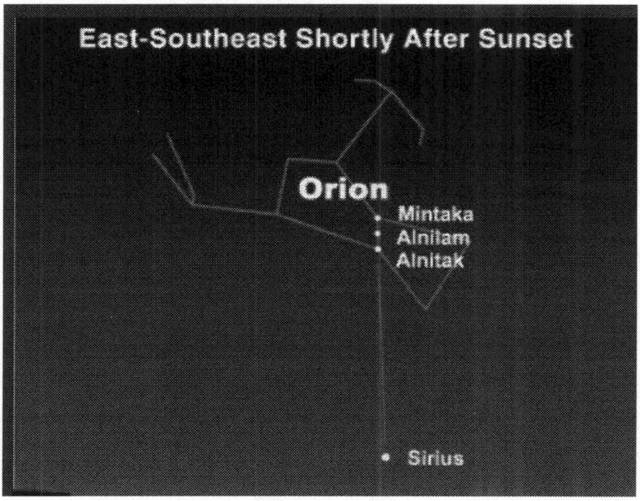

Now, if our Christian and Osirian mythologies are pointing to the literal Orion Belt and Sirius—and let us remind ourselves we are speculating rather wildly—what makes this more than enticing is that Sirius, is only 8.6 light-years from earth, far closer than the stars of the Orion Belt.

If our extraterrestrial visitors came from the region of Sirius, it would make their round-trip far shorter than from the Orion Belt.

The next question, thus, might be "are there any exoplanets orbiting Sirius?"

Sirius is in the constellation Canis Major and at least one rather large exoplanet has been identified which is some 67 light-years distant and in a habitable zone in this cluster.

But let us not lose sight of the Dogon either who mysteriously ceremonialized a companion star of Sirius, Sirius B, and believed humanity's ancestors came from there. That companion star was not discovered until 1862, [232] not photographed until 1970, but the Dogon, without telescopes, ritualized its existence since AD 1300. [233]

A round trip ticket at 80% of the speed of light to Sirius B would take just under 30 years.

Pi and phi

Beyond papal miters, elongated skulls, and Osiris, it is time for a discussion of ancient mathematics. We were taught that Archimedes in the third century BC discovered *Pi* and calculated it correctly at 3.14. However, if one calculates the height of the Great Pyramid at Giza, which was built far earlier than Archimedes, and multiplies that by 2*pi*, the exact perimeter of its base is defined, hardly a mathematical accident. Others dispute these interpretations.

$$\pi = \frac{C}{d}$$

When one takes a calculator to Mexico to do the same with the Pyramid of the Sun, the height of the pyramid X 4 *pi* equals the perimeter of its base, or 2,932.8 feet. That calculation is *less than half an inch* from its true perimeter. [234]

Another mathematical study of *Pi* in the Great Pyramid is given by professor Assem Deif from Cairo University:

---

[232] Robert Olsen, Can the Sirius star system support life, Owlcation, Dec 30, 2023, https://owlcation.com/stem/Can-The-Sirius-Star-System-Support-Life

[233] Lee Krystek, "The Dogon, the Nommos, and Sirius B," No Date given, http://www.unmuseum.org/siriusb.htm

[234] Graham Hancock, "The Sun and the Moon and the Way of the Dead," Fingerprints of the Gods, Crown Publishing, 2000 http://www.bibliotecapleyades.net/egipto/fingerprintgods/fingerprintgods06.htm

> To work out an Egyptian value for *Pi* from the dimensions of the Great Pyramid, they had a unit of length called the Royal Cubit (about 0.524m). By transforming the pyramid's height and base from western units (feet or meters) into cubits, it becomes evident that the designed height measures 280 cubits and the base 440.... By dividing half the base by the height (cotangent of the slope angle) one reaches the ratio *Pi*/4. It then follows that the Egyptian value for *Pi* is 3.142857143 being equal exactly to 22/7.[235]

What is peculiar here is that the Pyramid's value of *Pi* is accurate to two digits 3.141 vs. 3.142 (above). Archimedes calculated it to two digits more than two millennia *later.*

How did the Pyramid builders achieve such precision in 2,500 BC?

I wrote to Dr. Deif, who referred me to another Egyptian authority. I asked this question in my email: "How did the Pyramid builders achieve such precision in 2,500 BC?" His answer was interesting:

> Some think that it was just a sheer coincidence to have the pyramid so harmonically proportioned. But we have the records of Herodotus 2,500 years ago telling us that the Egyptian priests told him that this pyramid was intentionally designed so that the area of each face was equal to the square of its height. Herodotus' reports are further corroborated by the actual dimensions in the Ancient Egyptian units of cubits: 280 for the original height and 440 for the side of the base. The ratio of these two figures [280/220 = h/b = 14/11] corresponds to the square root of the Neb (Golden) Proportion, as follows:
>
> Height / ½ Base = 280 /220 = 14/11 = 1.272727
>
> Square of 1.272727 = 1.619 = Golden Ratio
>
> There is a surprising bonus. Divide twice the base by the height and you get 3.14, a practically perfect value of the Circle Index.
>
> 2 Base / Height = 880/ 280 = 3.1429 = Circle Index
>
> 3. 14159 – 3.1429 = 0.0013
>
> Difference = 0.04%

[235] Assem Deif, "Last of the ancient wonders," Research Gate, Mar. 2008 https://www.researchgate.net/publication/267447516_Last_of_the_Ancient_Wonders_III_Pi_Phi_and_the_Great_Pyramid

> Incorporating the Circle Index *(Pi)* into the Pyramid's design is also significant. The angle of elevation of 51o 50' 35", expresses the Circle Index (22/7) with very considerable precision. The angle of ascent gives the pyramid a unique geometrical property that represents the mystical squaring of the circle: that the ratio of the pyramid's perimeter to its height is equal to double the Circle Index...This, however, is not the first time that the Ancient Egyptians used these important relationships, because Snefru's Pyramid at Meidum has the same geometric characteristics as Khufu's (Cheops') Pyramid. Incorporating both these sacred ratios into the design of Egyptian buildings was no coincidence. All Egyptian temple gateways were designed to incorporate both *phi* and *pi*—thousands of years before the Greeks.' [236]

This comment does seem to add corroboration to the idea that the ancient Egyptians might have had at their disposal a mathematics that modern man did not discern for another two thousand years.

In an attempt to add more weight to Prof. Deif's assessment, a second academic study argues that the value of *Pi* in the Great Pyramid was *accurate to six digits,* [237] while Archimedes value was correct to only two; [238] so it would appear that the architects of the Great Pyramid were not *accidentally* ahead of their time.

As we learned in high school geometry, Pythagoras was credited with the discovery of *Phi* (Φ) an irrational number called the Golden Section, also referred to as the Golden Triangle, or the Divine Proportion. Pythagoras lived in the fifth century BC, but two millennia prior to that *Phi* is found in the construction of the King's Chamber of the Great Pyramid at Giza. [239] And it was also found in Luxor. [240] [241]

A charming contemporary account of *Phi* is given in the *The DaVinci Code.* Here is an excerpt:

> Nobody understood better than DaVinci the divine structure of the human body. DaVinci actually exhumed corpses to measure the exact proportion of human bone structure. He was the first to show

[236] personal email

[237] Hermann Schelper, "The chronology of pi." 2000, Springer, http://link.springer.com/chapter/10.1007/978-1-4757-3240-5_35

[238] "Why is Pi actually important," https://www.angio.net/pi/whypi.html

[239] cited in Hancock, Ibid. p. 336.

[240] Archeologist Schwaller de Lubicz also was able to demonstrate that the Golden Section was "applied at Luxor and with a complexity and sophistication never achieved by the Greeks." "Pushing back the portals of civilization, J. Douglas Kenyon, in Forbidden History, Ibid., p. 89 Further examples of the Golden Section in the Great Pyramid can be found at Kelly, Paul R.1 Mathematics Teacher; Oct2004, Vol. 98 Issue 3, p206-207, 2p

[241] Lucy Laney, "Karnak," http://www.ralph-abraham.org/courses/math181/math181.S96/lectures/lecture.5w/gold-egypt/karnak.html

the human body is literally made of building blocks whose proportional ratios always equal *PHI*....Try it. Measure the distance from the tip of your head to the floor. Then divide that by the distance from your belly button to the floor. Guess what number you get?

Not *PHI!* one of the jocks blurted out in disbelief.

Yes, *PHI*, Langdon replied. "One-point-six-one-eight. Want another example? Measure the distance from your shoulder to your fingertips, and then divide it by the distance from your elbow to your fingertips. *PHI* again. Another? Hip to floor divided by knee to floor. *PHI* again. Finger joints. Toes, Spinal divisions. *PHI. PHI. PHI.* My friends, each of you is a walking tribute to the Divine Proportion.

...*PHI* appeared in the organization structures of Mozart's sonatas, Beethoven's Fifth Symphony, as well as the works of Bartok, Debussy, and Schubert... In closing, Langdon said, walking to the chalkboard, "We return to symbols." He drew five intersecting lines that formed a five-pointed star. "This symbol is one of the most powerful images you will see this term. Formally known as a pentagram—or pentacle, as the ancients called it—this symbol is considered both divine and magical by many cultures...because if you draw a pentagram, the lines automatically divide themselves into segments according to the Divine Proportion...the ratios of line segments in a pentacle all equal *PHI*, making this symbol the ultimate expression of the Divine Proportion. For this reason, the five-pointed star has always been the symbol for beauty and perfection..."[242]

Phi is discussed in greater depth in Giza:

The Great Pyramid has a base of 230.4 meters (755.9 feet) and an estimated original height of 146.5 meters (480.6 feet). This also creates a height to base ratio of 0.636, which indicates it is indeed a Golden Triangles, at least to within three significant decimal places of accuracy. If the base is indeed exactly 230.4 meters then a perfect golden ratio would have a height of 146.5367. This varies from the estimated actual dimensions of the Great Pyramid by only 0.0367 meters (1.4 inches) or 0.025%, which could be just a measurement or rounding difference.[243]

Of course there is always dispute over these calculations, but to mathematician Eckhart Schmitz,

[242] Dan Brown,The Da Vinci Code, pp. 95-96. Doubleday, 2003

[243] Gary Meisner, "Phi, Pi, and the Great Pyramid of Egypt," 18 Aug, 2012, The Golden Number, http://www.goldennumber.net/phi-pi-great-pyramid-egypt/

> It is quite clearly evident that the builders of the Great Pyramid intentionally chose a perimeter measurement that would relate to 36 degrees Geodetic Latitude and in so doing expressed, not only a highly precise knowledge of the Earth's size and shape, but also demonstrated a precise knowledge of the value of the Golden Ratio or Phi the value of Pi and their relationship to a spherical body.[244]

It gets worse (again!)

$$\Phi = \frac{\sqrt{5} + 1}{2}$$

Mathematician Mark Reynolds then adds more substance to these speculations and relates the geometry of the Great Pyramid to the Pyramid of the Sun in Mexico.

> Out of all the number systems and numerical relationships that could have been used, the Egyptians selected the golden section ratio and its square root in the building of the Great Pyramid. The Pyramid of the Sun also contains a sophisticated and subtle "infusion by association" of the same ratio....That two different architects would use points in the geometric constructions in the singularly unique way I have discussed seems, to me, most improbable, perhaps impossible, for they serve neither a functional purpose, being totally unnecessary for construction purposes, nor a specific spiritual, symbolic, or philosophical purpose. However, it appears intentional. And if it was intentional, was it also intentionally mysterious?[245]

Reynolds work appears in a peer-reviewed mathematics journal and he may be the one of the more definitive authorities on this subject.[246] Beyond confirming that both *Pi* and *Phi* are related to Teotichuhan *and* Giza, he found other relationships that defy our understanding. He asserts, for example, that the Great Pyramid encodes information about the sun and the moon:

[244]Ekhart Schmitz, The Great Pyramid of Giza, Adventure Unlimited, 2003, http://www.thegreatpyramidofgiza.ca/content/index.html

[245]Mark Reynolds, "A comparative geometric analysis of the heights and bases of the Great Pyramid of Khufu and the Pyramid of the Sun atTeotihuacan,1999, Nexus Network Journal, https://www.emis.de/journals/NNJ/Reynolds.html

[246]Ibid.

> If the base of the Great Pyramid is equated with the diameter of the earth, then the radius of the moon can be generated by subtracting the radius of the earth from the height of the pyramid. [247]

Even more mind-boggling elements of the Great Pyramid are delineated by Livio Catullo Stecchini, an American professor of the history of science and an acknowledged maven on ancient measurement; he says the evidence for the existence of such anomalous knowledge in antiquity is irrefutable.[248] [249] He says the Great Pyramid of Giza "is essentially a scale model of Earth and a geodesic marker of the center of Earth's landmass. The ratio of the pyramid's height to its perimeter equals the ratio of Earth's radius to its circumference."[250]

The traditional notion that Pythagoras discovered *Phi* now seems untrue. It was incorporated into one of the earliest structures made by man, far earlier than the time of Pythagoras, and can be found in its rather glorious exactness in Giza.

More examples of human precocity

Instead of engaging in the hubris implicit in thinking our ancestors were backward, we might also reflect on the computation of the solar year.

The Gregorian calendar computed the exact length of the solar year as 365.2422 days, incorporating an error of just 0.0003 of a day. That was the pinnacle of scientific knowledge in the sixteenth century. But the Mayan calendar achieved greater accuracy far earlier than that computing the solar year with an error of only 0.0002 of a day.[251]

We are not really at the mercy of Van Daniken and the tabloid press on these matters. Knowledge of *Pi* and *Phi* reaches far back into antiquity along with a scrupulously and exacting astronomy and mathematics.

> Indeed, at the time of its construction, the height of the Great Pyramid was said to be 147 meters, and the distance of the earth from the sun at its closest point is 147 million kilometers. [252] Multiply the

[247] Ibid.

[248] Hancock, Ibid, pp. 179-180.

[249] It is also said by the Third century AD Chinese mathematicians could compute the value of pi, and by the fifth century they calculated it to ten decimal places..."a feat unmatched in the West until the sixteenth century." Feats and Wisdom of the Ancients: Time-Life, Alexandria, Virginia, 1990.

[250] Will Hart. The Genesis Race. Rochester VT: Bear & Co.,2003, p77

[251] The Mayan calendar put the exact length of the year at 365.2420 days. The modern Gregorian calendar puts it at 365.2425 days while the actual duration is 365.2422, making the Mayan calendar more accurate than the one commonly used today. Feats and Wisdom of the Ancients: Time-Life, Alexandria, Virginia, 1990.

[252] Peter Kosimo, Not of this World, Citadel Press, 1988,Note there is some controversy on these calculations. The Great Pyramid's facing stones have been removed over time, so

height of the pyramid by one billion, and you get the distance of the earth to the sun at the perihelion. The distance from the Earth to the sun was not known or calculated until 1860, yet is it possible the architects of the Great Pyramid suffused another little tidbit of astrophysics into their monument as well?

Maybe we can dismiss these facts as lucky coincidences, but a competing point of view is that 4,400 years ago, a civilization of incredible brilliance existed which was one of the very *first* recorded civilizations our species produced. And let us not forget that in its mythology, *it attributed its genius, its origin, and even its language to non-earthly sources, star people, or what we today call deities, Osiris, Horus et. al— or should we say Amen!*" [253]

Knowledge of the solar system?

There is a Sumerian clay tablet which is some 4,500 years of age that portrays the sun surrounded by nine planets. [254] It could be that this is merely an artistic drawing, but is it entirely ridiculous to suppose this might be an accurate description of our own nine-planet solar system rotating about the sun?

If true, it would mean that the heliocentric theory of Copernicus five hundred years ago was antedated by 4,000 years.[255] And Neptune, the ninth planet — not discovered until 1930—was in human awareness four millennia prematurely.[256]

---

its actual height at construction is an estimate. According to Alaa K. Ashmawy, author of *Seven Wonders of the World*, the height of the Great Pyramid is 146.75 meters. The exact distance to the Sun at perihelion is 147.5 million kilometers. So if one continues to use 1 billion as the multiplier, then multiplying the height of the Great Pyramid gets the perihelion with a margin of error of one half of one percent. In other words, either this is a very lucky coincidence, or the Egyptians built a pyramid encrypting in its structure subtle astronomical and mathematical facts, one of which was that the height of the pyramid predicted the closest distance to the sun with an accuracy of 99.95 percent! Also Peter Thompson's Secrets of the Pyramid, where the limestone covering on the Great Pyramid which was stripped by vandals was used to estimate the actual height of the pyramid. Although Thompson does not discuss the perihelion, he does argue that its actual height was 147.9 meters which is consistent with this allegation. See Peter Thompson, *Secrets of the Great Pyramid*, New York: Galahad books, 1971.

[253] Jerry Kroth, *Aliens and Man*, New York, Algora, Nov. 2010.

[254] Another example of precocity is in Hindu history. It is said that the ancient Hindus knew of the seventh planet, Uranus, long prior to Herschel's discovery of it in AD 1781. See *Gods and Spacemen in the Ancient East*, p. 29, W. Raymond Drake. New York: New American Library, 1973.

[255] And if not, then it is clear that many had the idea of the earth moving around the sun. Aristarchus of Samos, 1700 years prior to Copernicus, made this discovery as well. *Feats and Wisdom of the Ancients*, Time-Life Books, Alexandria, Virginia, 1990.

[256] In India, "Kananda and the Gnani Yogis speculated on the atom five hundred years before Democritus, Aryabatha in the sixth century before Christ taught the rotation of

Curiously, in 2,200 BC Hindus and Assyrians represented Saturn with a ring of serpents, but no other planets were given rings.[257] Were these symbolic serpents again just lucky coincidences? If not, how did Hindus, two millennia before Christ, come to know Saturn had rings?

More mysteries

Longitude, we were taught, was mastered in the 1720s, but the map of Oronteus Fineas appear to disclose knowledge of longitude and the capacity to do spherical trigonometry, the mathematics involved in computing projections on a globe. [258] One Russian scientist even opines that:

> between the 5th and 10th millennium BC, there was a civilization on Earth that possessed great knowledge in the field of navigation, cartography, and astronomy, which was no less advanced than that of the 18th century."[259]

the Earth, the scientific principles of medicine, botany and chemistry were established as early as 1300 BC in India. Drake, Ibid., p. 63.

[257] W. Raymond Drake. *Gods and Spacemen of the Ancient Past*, New York: New American Library, 1974.

[258] Graham Hancock, Ibid., p. 30.

[259] As an illustration of the vast treasures of knowledge lost to antiquity, it is said that the Egyptian Library in Alexandria in 323 BC possessed ten times the number of volumes that the entirety of Europe possessed in 1450AD. The Library in Alexandria was burned in 46BC, again in 391 AD and finished off by Arab legions in 640AD.

Could it be possible some kind of advanced society existed prior to this time antedating our knowledge of our own civilization and one skilled in mathematics, cartography, and astronomy? Could it have spread its influences to the Egyptians and the Chinese as well as to have mapped Antarctica? It occurred to me that rather than reflexively resorting to the ancient-alien hypothesis, *it still could have been just a human society*. A good example might be mythic Atlantis as we have discussed earlier.

Many of these incongruities could be explained by simply postulating an advanced human society, which we have no records of, rather than resorting to the alien explanation.

But it is also possible to proffer an equally speculative suggestion, namely that there could have been some extraterrestrial involvement or interaction with humankind which helped to produce these anomalies as well. It is no less implausible to suggest *that* than to attribute it all to an Atlantis we have never found.

Baalbek, Peru, and Bolivia

There are a few "lesser" mysteries that also baffle our preconceived notions of a backward, undeveloped ancient world. Not only are the stone-cutting techniques notable in the construction of the Egyptian pyramids, but one giant carved stone in the Middle East at Baalbek is 1,100 tons, 72 feet long, and the largest piece of hewn rock on the face of the earth. This one, single, sculpted rock weighs in at 2.4 million pounds! To budge it one inch would require an army of more than 16,000 workers. The enormous crawler which is currently used by NASA to move the Saturn V rocket is thought to be the only machine on earth today which could move this monolith.

In Sacsayhuaman, above Cuzco, Peru, there are similar unwieldy megaliths perfectly cut, carried, and placed in strange arrangements. Some of these blocks weigh 300,000 pounds and are fitted together perfectly. The enormous stone blocks are cut, faced, and fitted so well that even today one cannot slip the blade of a knife or even a piece of paper between them. No mortar is used, and no two blocks are alike...Each individual stone had to have been planned well in advance, a 20-ton stone, let alone one weighing 80 to 200 tons, cannot just be dropped casually into position with any hope of attaining that kind of accuracy. The stones are locked and dove-tailed into position, making them earthquake-proof.

These megaliths not only are a mystery of construction but the entire complex is as well. The stones are organized to cut an angle of six degrees so that when the winter solstice occurs, the resulting shadows are cast to fall precisely at the foot of the protruding walls. The complex is also an acoustic marvel with echoes focused precisely on a central point across a plaza the size of a football field.

150-ton stone, Baalbek, in Lebanon

Much thinner than presently, I'm standing at Sacsayhuaman, Peru.

Indeed, after many devastating earthquakes in the Andes over the last few hundred years, the blocks are still perfectly fitted while the Spanish Cathedral in Cuzco has been leveled twice. Even more incredibly, the blocks are not local stone, but by some reports came from quarries in Ecuador, almost 1,500 miles away. Others have located quarries a good deal closer, only five miles or so away.

Though this fantastic fortress was supposedly built just a few hundred years ago by the Incas, they have no record of having built it, nor does it figure in any of their legends. How is it that the Incas, who reportedly had no knowledge of higher mathematics, no written language, no iron tools, and did not even use the wheel, are credited with having built this cyclopean complex of walls and buildings? Frankly, one must literally grope for an explanation, and it is not an easy one.[260]

Another investigator tried to learn the origin of the rocks from a nearby quarry to Cuzco. He made the following observation:

> The largest stone...weighs about 150 tons. It could have been pulled up a ramp with a force of about 260,000 pounds...such a feat would have required a minimum of some 2,400 men. Getting the men seemed possible, but where did they all stand? ...Further perplexing ... is that the stones of Sacsayhuaman were finely dressed, yet are not polished, showing no signs of dragging. He could not figure out how they were transported the 22 miles from the Rumiqolqa quarry.[261]

The stones of Baalbek and Cuzco do not show signs of having been dragged. How were they cut, moved, or placed? And why were such large stones used rather than cut into more manageable pieces? According to author David Childress:

> If they had cut the stone into, say, 100 pieces, they would still be of unusually large size, larger than a man, but at least could have been stacked into a wall much more easily. One is left with the unsettling thought that the reason they used these huge stones was because they *could* use them—and do it relatively easily, though today we have no idea how.[262]

Anthropologists describe the Neolithic period (literally 'new stone age') as one of hunter-gatherers who still have no written language, trap fish, make beads, and are only in the early stages of domesticating plants. But appearing right alongside these primitive peoples, we find inscrutable megaliths and pyramids built using mathematics, astronomy, and intricate construction technologies which absolutely mystify us.

Recently scientists discovered a large pyramid and temple-observatory in Peru antedating by 800 years all archeological assessments of when advanced civilization took root in the Americas. The ancient observatory precisely predicts the solstices and was built *within 400 years of Giza, over 7,000 miles away.*

> "It is really quite a shock to everyone...to see sculptures of that sophistication coming out of a building of that time period," says Yale

[260] David Childress, *Technology of the Gods*, Kempton, Illinois, Adventures Unlimited Press, 2000, pp. 51-52.
[261] Childress, Ibid., p. 57.
[262] Childress, Ibid., p. 45.

> archeologist Richard Burger [263] ...Even the name of the culture which inhabited the region remains unknown, because "writing did not emerge in the Americas for 2,000 more years." [264]

90 degree angles cut in stone (Egypt, left) and Peru (right)

Subjective observations

In preparation for writing this book, I traveled to Egypt to visit Delphi, walked the Nazca lines, explored Cuzco and Sacsayhuaman, and visited the White Pyramid in China. One observation stuck with me: When I was in Egypt, I noticed a 90-degree angle cut in stone, and it amazed me, but when I got to Peru, I found the same kind of stone work. Others have noticed this too.[265]

And yet despite the similarities, we are taught that these cultures, 7,000 miles distant from each other, had no connection with one another. It causes one to wonder.

A footnote about scientific humility and speculation

A thoroughly enjoyable scientific read is Bill Bryson's *A Short History of Nearly Everything*. It reminds us of how flimsy our scientific knowledge is. Take the emergence of humankind. Bryson says that fossils are extremely difficult to form and very susceptible to degradation over time. Those few we have discovered from Australopithecus to Lucy to Neanderthal to the earliest *Homo sapiens*—that is, the rare fossilized bones and fragments from which we have reconstructed *the entire history of mankind*—could all fit in the back of a pick-up truck.

In other words, all the raw empirical evidence used to reconstruct our five million-year journey from chimps through all the hominid variants to our present species can fit in the back of a Chevy. Kind of makes you cautious in attaching high levels of confidence to present scientific formulations about mankind's evolution and origins.

Mainstream archeology says that migrations occurred into the Americas only about 11,000 years ago. [266] Oops, a few years after that pronouncement, an archeologist reported that fossilized human excrement found in

[263] "Celestial find at ancient Andes site." Thomas Maugh, *Los Angeles Times*, May 14, 2006, p. A28.

[264] Ibid.

[265] "The connection between Peru and Egypt," NephiCode, 0 March, 2013 http://nephicode.blogspot.com/2013/03/the-connection-between-peru-and-egypt.html

[266] Haynes, Gary (2002). *The Early Settlement of North America: The Clovis Era.* New York: Cambridge University Press.

Oregon carbon-dated to 14,300 years. [267] Well, okay, then, 14,300 years ago there were migrations across the Bering Strait. But, oops again, a 2009 study analyzed the dispersion of genes in the Americas and pushed back the date to 17,000 years.[268] A 2016 report says humans reached Chile by 18,500 years ago.[269] A recent DNA study pushes the date back to 20,000 years. [270] But a 2017 study found humans were in North America 24,000 years ago.[271] Then, to top it off, Professor Silvia Gonzalez of John Moore's University discovered a human footprint in volcanic ash in central Mexico carbon dated to 40,000 years.[272] All of this seems to happen rather regularly.

So maybe *Homo sapiens* made it to North and South America only 11,000 years ago, or took a tell-tale dump in Oregon 14,300 years ago, or had genes which pegged his beginnings here at 17,000 years, or perhaps 20,000 years, and then again maybe he wasn't wearing shoes 40,000 years ago, as Dr. Gonzalez says. [273]

Then, is it really completely out of the ballpark to think that ancient Egypt, Peru, Teotichuhan, and even China may have been connected in antiquity in ways we have yet to understand, and that our knowledge of ancient times is not quite as solid and unshakable as we have been taught?

[267] "14,300 years old fossilized feces suggests earliest human presences in North America." *Science News*, April 3, 2008.

[268] Bruce Bower, "Migrants settle Americas in tandem." *Science News*, Jan. 31, 2009, p. 5.

[269] Bruce Bower, "Arrival to South American pushed back," *Science News*, Dec. 26, 2016

[270] P. Forester, et. al., "The origin and evolution of Native American mtDNA variation: a reappraisal," Am J Hum Genet. 1996 Oct; 59(4): 935–945. https://www.ncbi.nlm.nih.gov/pmc/articles/PMC1914796/

[271] University of Montreal, "First humans arrived in North America 10,000 years earlier than thought," 16 Jan, 2017, Phys.Org. https://phys.org/news/2017-01-humans-north-america-years-earlier.html

[272] Sources: "Alleged 40,000-Year-Old Human Footprints In Mexico Much, Much Older Than Thought," *Science Daily* (Dec. 1, 2005); see http://www.sciencedaily.com/releases/2005/11/051130232517.htm

[273] University of South Carolina, "New evidence puts Man in North America 50,000 years ago, Nov. 18, 2004. There are even speculations that *Homo Sapiens* might have been in the Americas 50,000 year ago or even 130,000 years ago. See https://www.sciencedaily.com/releases/2004/11/041118104010.htm see also Gemma Tarlach, "The first Americans may have arrived 130,000 years ago."16 Apr. 2017. Discover https://www.discovermagazine.com/planet-earth/the-first-americans-may-have-arrived-130-000-years-ago

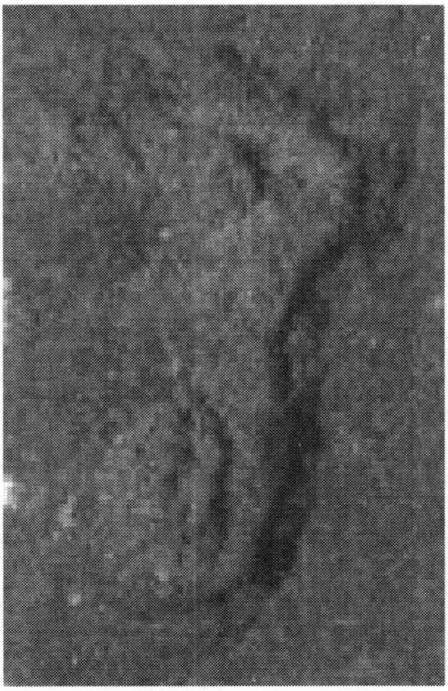

40,000 year-old human footprint found in Mexico

Tough talk from the mainstream

Some archeologists sit up on their haunches with palpable "disgust" over these speculations that ancient Egypt might be connected to Teotichuahan, or Xian, or Thornborough or Peru. In their din and roar they speak defiantly about such "fantasies" created by what they call "diffusionists."

To archeologist John Howland Rowe,

> Doctrinaire diffusionism is a hardy weed, however, and no sooner had it been poisoned in ethnological pastures than it crossed the fence and began to infest archeology.[274]

Clearly, the mainstream has quite another opinion about all these matters and doesn't take kindly to such inquiries considering them rather poisonous *infestations* of the purity of their discipline.

[274] John Rowe, "Diffusionism and Archeology," American Antiquity, Jan. 1966, Cambridge University Press http://www.jstor.org/stable/2694735?seq=1#page_scan_tab_contents

The Voynich Manuscript

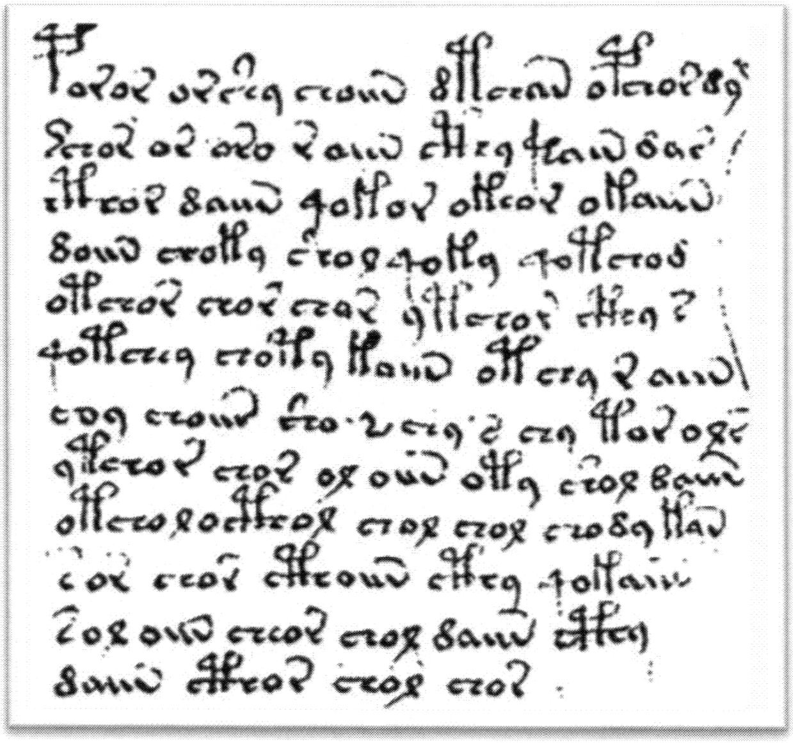

Undaunted, however, we will proceed on our archeological museum tour of possible extraterrestrial exhibits to our next room which houses the mysterious Voynich manuscript. This is a 235-page book which takes its name after American antique book dealer Wilfrid Voynich. It was found in 1912 among a collection of ancient manuscripts kept in a villa near Rome. The manuscript has 20 pages missing and is estimated to have been created in the late 13th century. It contains text and drawings of many unidentified plants, what appear to be herbal recipes, mysterious charts, nudes, and astronomical objects.

The book is written in a language which has an alphabetic script (see right), and is thought to contain between 19 and 28 letters . The manuscript has been carbon dated to 1404-1438.[275]

[275]Stolte, Daniel, "Experts determine age of book 'nobody can read,' PhyOrg, Feb, 2011.

This syllabary bears no relationship to any English or European lettering system.

The earliest date to which this text can be fixed is 1586. Emperor Rudolph II in Bohemia purchased it at that time. A letter accompanying the manuscript said that it was the work of Englishman Roger Bacon, who lived in the 13th century.

> William F. Friedman was an Army cryptographer, who was among the first people to use computers for textual analysis. In 1925, Friedman and his wife, Elizabeth, also a cryptographer, worked on the manuscript project for forty years. Friedman and his colleagues broke Japan's code Purple during the Second World War, and Friedman became the chief cryptanalyst for the War Department and head of the Signals Intelligence Service in the forties and fifties. The historian David Kahn called him the "world's greatest cryptologist." By 1944, Friedman had formed the Voynich Manuscript Study Group with some colleagues. The group never cracked the code." [276]

Virtually every attempt to decipher it has also failed. Decoding attempts began in earnest after the advances made in cryptography. After World War II, the National Security Agency in Washington sponsored a symposium on it in 1978 but failed to decipher it. [277]

Some computer findings are of interest. For example, one element of professional cryptography known as "Zipf's law" provides statistical criteria to determine if a language or code is a hoax. The Voynich manuscript conforms to standard word frequencies, word repetitions, and appropriate distances between letters and words. These internal consistencies suggest that it is not a hoax but is written in a natural, albeit unknown, language.

Some researchers say that the biological drawings in the text described "asseminiferous tubes, microscopic cells with nuclei, and even spermatozoa," items hardly known at the time the book was written.

Others have called the language and its script "Voynichese." The letters loop prettily, and the text runs from left to right, top to bottom. The first half of the book is filled with drawings of plants; scholars call this the "herbal"

[276] Jo Livingston, "The unsolvable mysteries of the Voynich manuscript," The New Yorker, 30 Nov. 2016. http://www.newyorker.com/books/page-turner/the-unsolvable-mysteries-of-the-voynich-manuscript

[277] John Titlman, "The most mysterious manuscript in the world," 21 Apr. 2002https://www.nsa.gov/news-features/declassified-documents/tech-journals/assets/files/voynich-manuscript-mysterious.pdf

section. None of the plants appear to be real, although they are made from the usual stuff (green leaves, roots and so on.)[278]

Two astronomical drawings are alleged to have been deciphered, descriptions of spiral nebulae and a coronal eclipse.

The manuscript is heralded as the most studied and still the *most mysterious* manuscript in the world. Hundreds of researchers examined it all with no ultimate resolution. There is an on-going effort on the internet to decode it. Presently, all that is known about it is that it is *not related to any known language or any linguistic derivative.* Attempts to link it with Ukrainian, Latin, numerical systems, or European languages have all fallen on their face. One study said that the script appears to be written either by two separate individuals or by one individual at two different parts of his life. There are always alleged "breakthroughs" reported, some saying the text is in archaic "phonetic" form of Turkish. But despite the proliferation of such "discoveries" no one has translated more than a hypothetical portion and no one has successfully explained the plants and root diagrams."[279]

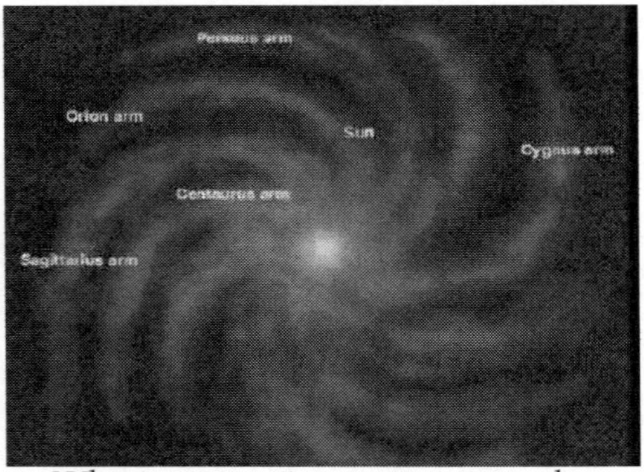

What is most curious to our interest, however, is that one of the diagrams has been likened to a crude drawing of the Milky Way. And in the graphic which follows, the superimposition of one on the other is impressive. So this is either one more lucky coincidence or the author had an uncanny foreknowledge far before any human beings came into possession of such information. [280]

[278] Jo Livingston, "The unsolvable mystery of the Voynich Manuscript," *The New Yorker*, 30 Nov. 2016. http://www.newyorker.com/books/page-turner/the-unsolvable-mysteries-of-the-voynich-manuscript

[279] "Has the Voynich Manuscript finally been decoded," 19 Feb 2019, Open Culture. https://www.openculture.com/2019/02/has-the-voynich-manuscript-finally-been-decoded.html

[280] The spiral arms of the Milky Way were only discovered in the 1950s; *Matson, John (September 14, 2011).* "Star-Crossed: Milky Way's Spiral Shape May Result from a Smaller Galaxy's Impact". *Scientific American.*

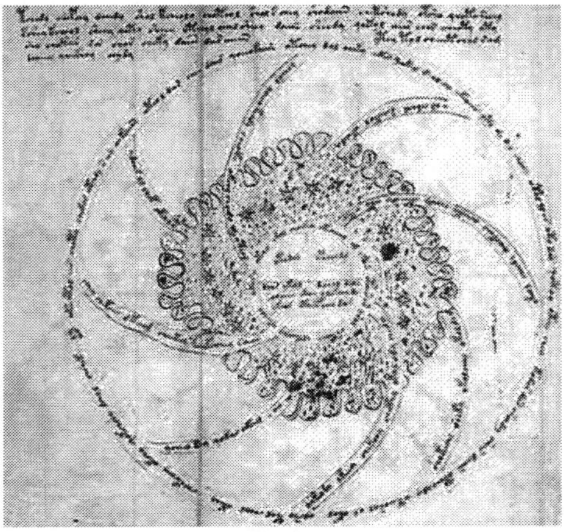

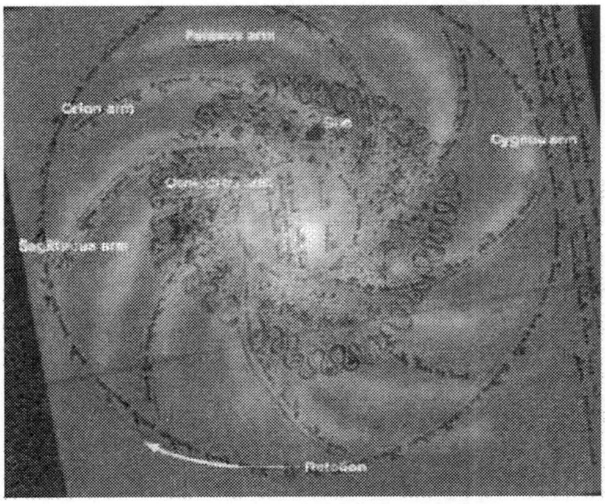

See the diagram of the Milky Way in http://en.wikipedia.org/wiki/Milky_Way

The Voynich manuscript contains a drawing one author likened to a crude depiction of the Milky Way spiral galaxy. When the spiral arms of the galaxy are superimposed on the drawing, it has an uncanny similarity. Author's note: The Voynich manuscript seems to depict 8 arms, (upper left); however, current astronomical science believes there to be only four. However, if one considers the four arms in rotation, (lower left) they give the appearance of eight. Perhaps the Voynich manuscript is the work of a highly intelligent schizophrenic who created his own language, script, and alphabet. Schizophrenics are known to create fragmentary languages, but one who could make a consistent alphabet, word frequencies, and respect

Zipf's law consistently for all 223 pages would mean the schizophrenic was extremely gifted.

It might also be the product of a secret alchemical sect, but if so, one would think 21st century CIA decrypters or the National Security Agency would have decoded it and linked it to some existing language or secret algorithm which would by now have given up its codes.

The most far-flung hypothesis, of course, is that the reason it has eluded modern cryptography's search for a language to which it might be linked is precisely because the manuscript *has no human linguistic counterpart,* that is, a language unknown on this earth.[281] In this sense, then, it represents an extraterrestrial artifact.

Summary: Archeology unexamined

Sego Canyon, Utah rock art between 2000-8000 years old; source: http://albertdebruijn.com/home/archives/2842

[281] A recent text reviews the major theories about the manuscript, including the extraterrestrial one. See Lawrence and Nancy Goldstone, The Friar and the Cipher, New York, Doubleday, 2005. See also The Voynich Manuscript: The Unsolved Riddle of an Extraordinary Book which Has Defied Interpretation for Centuries," by Gerry Kennedy and Rob Churchill.

We have incredibly accurate Mayan solar calendars, thousand year-old figurines wearing helmets and goggles, cross-cultural pyramid planes dotting the globe pointing to the Orion Belt and invoking sky-god deities—not to mention the Thornborough Henge— plus incredible numbers of rock paintings and petroglyphs depicting helmeted, big-eyed and big-headed humanoids with antennae, lizard-like bi-pedal beings described in mythology, pictographs showing representations of UFOs, Egyptian royalty with their elongated skulls and dolichocephalid Peruvian and Siberian counterparts, ancient maps of mountains and rivers in Antarctica which might have been drawn 3,300 years before Captain Cook arrived—and using spherical trigonometry to boot—not to mention one of the most mysterious manuscripts in the world which has eluded translation for 600 years and which may contain a prescient description of the spiral arms of the Milky Way half a millennium before modern humans made that discovery.

Add into this mix the staggering mathematical sophistication of the Pyramid builders in 2,500 BC, antedating by 2,000 years the discovery of *Pi* and skillfully using *Phi* both in Giza, Luxor, and in Teotichuahan.

At the end of Part I of this text, we asserted that if a U.S. Congressional hearing investigated the UFO phenomenon, it would conclude by the sheer weight of evidence presented, that extraterrestrial contacts have occurred. In that sense, all of the archeological data presented here would take on different coloring too: it would begin to fit together; we would begin the process of seeing our *true history* and the events that shaped human civilization and culture. The fragments, and these isolated bits and pieces would begin to form a mosaic, and the full picture of our ancient history would organize itself in a way we have never been taught and in a way we never conceived.

Our entire view of early human history and the beginnings of human civilization would be altered.

We would not disdainfully mock the idea that Osiris, Marduk or their contemporaries influenced, shaped, or even may have launched the ascent of humanity.

Instead, we must confront the reality of how mainstream archeology has deceived us. It has lured us into an intellectual abyss with its absolutist preconceptions. This is such a fundamental assumption they—and most of the rest of us—simply take it for granted. Their basic tenet, archeology's sclerotic shibboleth is that there are no gods, no space creatures, no aliens, no visitations, no extraterrestrial contacts. On that basis, they insist on explaining to us how all these mysterious things happened *without recourse to any of those concepts*:

> The pyramids *had to be built by man.*
>
> Here's how they did it [they are still trying to figure that one out. Another dissertation, please!]
>
> Here is why the petroglyphs in Niger have no similarities to those in Utah [still not very convincing].
>
> Here is why the pyramid alignments in Xian could not possibly be related to Egyptian pyramids, much less Thornborough. All are either based on egregious miscalculations or nothing but random unintended concordances [Another dissertation please!]

Numerous books and periodicals underscore that malignant and interminable bias: Cult Archeology, Archeological Fantasies, Alien Chic, Fringe Archeology, Bad Archeology, Debunking Encounter Myths, The Idiocy, Ancient Aliens Exposed, The Archeology News Network.

The myriad stories of gods from Osiris to Quezalcoatl are simply superstitious, irrational fables, human fantasy, delusions, and in no way connected to reality. Therefore the calculations are wrong; the correspondences are purely coincidental. The ninety-degree carved boulder angles in Egypt and Peru nothing more than the imaginings of untrained observers; those cultures had no knowledge of one another. "Period!"

The Egyptians could not have had *Pi;* the mainstream insists that knowledge came two millennia later. That Sumerian glyph with nine planets, "Nonsense!" "Preposterous." "More Pseudoarcheology." The Voynich manuscript the writings of a space alien? "Garbage!"

"Poisonous diffusionists!"

And so it goes. Any attempt to bring these concepts into mainstream academic thinking is reflexively dismissed as faulty scholarship, new age mysticism, or Discovery Channel gobbledegook. It is sent packing with the same indignation and outrage that is directed at any of the souls who have reported UFOs, or tried to report them—or lost their jobs because they did so.

They are thrown into the very same dungeon of chuckleheads, crackpots, nut cases, and the unhinged. If there is any revolution in our thinking, perhaps one day we will finally start to see the kind of research we need to reconstruct our distorted, denied, skewed, censored, and forbidden history.

The God of Denial is alive and vigilant in keeping the ivory tower, on which it is perched, cleansed and purged of all impurities, all apostates, and all believers that humankind may have ever encountered extraterrestrials in its history. Sad indeed.

In 1798, Napoleon Bonaparte embarked on a military expedition of Egypt. Like so many throughout history, he was drawn to the mystique of the Pharaohs. Napoleon was an influential force in the birth of Egyptology. Accompanying his troops were also engineers, surveyors, artists and archaeologists who were required to document the great ruins of the ancient Egyptian civilization. The inner sanctum of the Pyramids drew Napoleon in, like so many before him and since. After exploring the chambers of Khufu's Great Pyramid, Napoleon requested to be left alone in the King's Chamber. When he finally emerged, he is reported to have been extremely shaken and shocked by something within. When asked what had happened, he refused to discuss it and insisted that the incident never be spoken of again. A friend who visited Napoleon in his final days asked him to tell him the secret of that day. Napoleon was ready to speak, but then shook his head and declined. Enigmatically, he responded, "No, what's the use. You'd never believe me." [282]

[282] Mystery Traveler http://tibetkanagawa.blogspot.com/2006/06/napoleon-mystery-within-pyramid.html

EVOLUTIONARY BIOLOGY AND THE EXTRATERRESTRIAL [283]

> The Nephilim were on the earth in those days—and also afterward—when the sons of God went to the daughters of men and had children by them.
>
> —Genesis 6:1-5

There is a pervasive theme in human mythology and in its myriad creation myths that goes beyond mere visitations by other beings but speaks of Gods who mate with humans or who biologically altered our species. This theme is found in myths in China, Greece, Rome, Sumeria, Germany, and *Genesis.*

Karl Abraham's dictum that fairy tales and legends encode human collective memories may be a lot more than that. During the devastating "Spanish flu"[284] of 1918, a worldwide pandemic which took at least thirty million lives, a nursery rhyme appeared:

> I had a little bird
>
> And its name was Enza
>
> I opened the window
>
> And influEnza.

President Wilson, who almost succumbed to the disease, murmured this verse in the Oval Office as did so many others across the nation. More than half a million Americans died in the epidemic in a ravaging nine months, far exceeding American casualties in World War I. The rhyme not only memo-

[283] Portions of this chapter originally appeared in Jerry Kroth, Aliens and Man: a synopsis of facts and beliefs, New York, Algora, 2001 and are reprinted here with the permission of the publisher.

[284] Note: this flu actually originated in the United States but came to be known as the Spanish flu.

rializes the event, but its most elegant component is that *it actually hints that the disease was 'bird borne.'*

In 1918 viruses had not yet been isolated and weren't known until 1931. Current research on this plague took extracts of frozen tissue from the preserved cadavers of victims of the 1918 flu and showed it to be a variant of swine flu.[285]

What we now know is that Canadian geese migrating south excreted droppings on Midwestern farms, which were absorbed by pigs. The bird-borne virus then incubated, mutated, and was passed on to human beings, eventually infecting the entire world. The rhyme about the bird, *Enza,* not only recalls the event but provides a diagnostic clue. completely unknown in 1918, *that the pandemic was of avian origin as it indeed was.*

There are a few other mysteries I would like to introduce as we gradually enter this controversial chapter.

Kirlian photography and ch'i.

By placing a living object directly on a photographic plate and running an electric current through it, what appears looks remarkably like an aura. First discovered in 1937 and known as Kirlian photography, the aura was most easily observable from a fingerprint. The fingerprint-aura is generally colored blue, and alters its size [286] depending on various mood states. If a psychiatric patient is in a manic state, it was expansive and pink; if the patient is depressed, it contracts into a deeper bluish tint. The size, coloring, and character also vary with a woman's ovulatory cycle. [287]

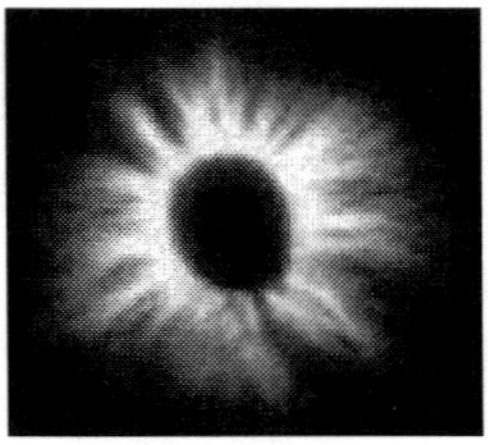

Kirlian fingerprint photo

What is uncanny is that prior to the discovery of Kirlian auras, *ninety-six* different cultures mentioned auras, halos, and egg-shaped bubbles of irides-

[285] Vana G, Westover KM (June 2008). "Origin of the 1918 Spanish influenza virus: a comparative genomic analysis". Molecular Phylogenetics and Evolution 47 (3): 1100–10.
[286] Photo source :http://interconnections.co.uk./Health/Kurlian
[287] Thelma Moss, The Body Electric, New York: Jeremy P. Tarcher Inc., 1979.

cence surrounding the human body. These descriptions can be found in the medieval writings of Christian mystics, the sacred works of ancient India describing *prana*, as well as China in 300 BC where the concept of *ch'i* is first mentioned. [288]

Even though science had not verifiably recorded *any such evidence until 1937*, the *prescience* of mythology antedated modern science by five thousand years.[289]

All of this creates a new respect for the ancient record.

Most scientists relegated the concept of auras and halos to the dustbin of superstitious folklore and never gave it a second thought. That same attitude holds true for much of what we will be discussing next, the possibility that the *Homo sapiens* genome has been tweaked in some manner through extraterrestrial interaction. Are these accountings mere superstitious, delusory attempts to give our species a divine veneer and a cosmic importance, or are these mythic reports of interaction between man and the Gods archaic memories?

Man as hybrid

The lyrics of ancient Greek (Orphic) song are: "I am a child of Earth and starry Heaven, but my race is of Heaven."[290] In Sumeria, the earth goddess Ki mates with the heavenly god, An, and from their intercourse comes Enlil.[291] Enlil, in turn, 1,500 years before the Old Testament, reports in the saga of *Gilgamesh* of the great flood and an ark that was built according to the specifications given by the gods.

In India the gods Shiva and Devi were involved in the creation of human beings. The royal family of Japan is said to have descended from an incestuous conception of the gods. In Ireland the goddess Danann gave birth to Bres, who was "given the kingship of Ireland."[292] In the Book of Noah and the Book of Enoch there are "explicit accounts of physical intercourse between the gods and humans."[293] In Christian mythology, God the Father mates with Mary and begets Jesus, half man, half God.

> Greek legends delight in the amorous revels of Zeus descending as a hero, bull, even a swan, to seduce any willing maid: Mars ravished the

[288] Roger Jahnke "History of the Chi Cultivation," Feel the Qi, no date given, http://www.feeltheqi.com/articles/rc-history.htm

[289] Michael Talbot. The Holographic Universe. Sep, 2011,Generic.

[290] Paul Von Ward, Gods, Genes, & Consciousness, Hampton Roads, 2004, p. 262.

[291] F. David Peat, Synchronicity: The Bridge Between Matter and Mind. New York, Bantam, 1987, p. 190.

[292] Paul Von Ward, Gods, Genes and Consciousness, Ibid, p. 23

[293] Ibid., p. 300.

> Vestal Virgin, Rhea Silvia to father Romulus; the Gods and Goddesses of Old India winged down for pleasure with mortals procreating Kings; the Babylonian Beauty Queen won a date with Marduk in that elegant boudoir topping his great Temple; the Japanese Emperors claimed descent from Amaterasu, Goddess of the Sun...Jehovah's visitation to the beauteous Sarah produced Isaac, father of Jacob...Such unions between Celestials and Earthmen and Earthwomen may not have been the lust we imagine but divine eugenics to create a new race.[294]

Just like 'ring around the rosy' points to the bubonic plague and the numerous flood myths may encode a memory of deglaciation from the last ice age—*that is, these myths seem to point to real things*—is it possible that such creation-interbreeding myths have a similar meaning and significance and represent archaic memories of actual events?

Evolutionary biology

We were taught that the process of evolution is gradual, that man came down a special evolutionary tree, and that the ultimate manifestation of *Homo sapiens* is a logical, and slow, process of evolution and natural selection. In elementary school we eagle-eyed drawings of apes walking behind a two-legged, upright hirsute Neanderthal, followed by other developing hominids until finally *Homo sapiens* appears as the endpoint of an evolution occurring over millions of years. (And sometimes he was even wearing a tie.)

In the last fifteen years, numerous findings impugn these traditional notions. One problem is that we seem to be far older than we ever thought. A second involves the recent appearance of a DNA anomaly called Mungo Man, and the final issue sucks us into the conundrum of how the evolution of human *culture* does not seem to advance in a way *paralleling* our physical evolution. We will drill down into these four topics in this chapter.

Neanderthal Kith and Kin?

For many years anthropologists were not convinced that we were descended from Neanderthal. It was thought he was merely a distant cousin: A Neanderthal bone discovered sometime ago was examined for retrievable DNA. Some was found, and it showed Neanderthal was not the forefather of *H. sapiens* and was genetically distinct.[295]

> The finding, by veteran ancient-DNA researcher Svante Paabo of Munich and his colleagues, could help bring down the curtain on

[294] W Raymond Drake, Gods and Spacemen of the Ancient Past. New York: New American Library, 1974, p. 98.

[295] "DNA sleuths redraw path of human development. " San Jose Mercury News, July 11, 1997, p. 22A.

> one of the longest-running and feud-prone disputes in anthropology: whether Neanderthals disappeared . . .because they gradually evolved into modern humans; or whether they were abruptly superseded by an evolutionary strain that left Africa only a few dozen millenniums ago and from which all people now alive are descended.[296]

It seemed then that *Homo erectus* was the common ancestor and that Neanderthals and modern humans evolved separately. Another study reports Neanderthal DNA differs from human and chimpanzee DNA in 27 different ways.[297] However, the most recent data suggests that indeed Neanderthal did mate with *Homo sapiens* and that individuals today incorporate Neanderthal DNA:

> Neanderthal-inherited genetic material is found in all non-African populations and was initially reported to comprise 1 to 4 percent of the genome. This fraction was later refined to 1.5 to 2.1 percent. It is estimated that 20% of Neanderthal DNA currently survives in modern humans.[298]

Neanderthal

[296] Ibid.

[297] That is to say, out of 378 nucleotides of DNA examined, Neanderthal mitochondrial DNA differed from that of humans and chimpanzees in 27 places. Human DNA is 98.4% identical to the DNA of chimps and bonobos. Orangutans, gorillas, chimps and bonobos all were left-brained while monkeys were not. "See Monkeying around with the brain. San Francisco Chronicle, May 11, 1998, p. A2.

[298] https://en.wikipedia.org/wiki/Neanderthal_genetics

Recent findings from Indonesia in 1996 show that *Homo erectus* was around as late as 53,000 years ago. This means that *Homo erectus* and *Homo sapiens* co-existed. One researcher alleges all three co-existed:

> It now seems that three separate human species—Neanderthals, anatomically modern humans, and an older species called *Homo erectus*—apparently coexisted until as recently as 30,000 years ago in part of Europe and Asia.[299]

A "hybrid" four-year-old boy was found in 1999 that was approximately 24,500 years old. He was part Neanderthal and part early modern *Homo sapiens.* The find pushes Neanderthal existence ahead 4,000 years.

(By the way, Neanderthal actually was taller and had a larger brain than *H. sapiens* by about 100 cc).[300]

More recent research comes from a 2009 study from the Max Planck Institute for Evolutionary Anthropology. They decoded 3.7 billion bases of Neanderthal DNA from a female bone found in Croatia. One interesting finding was that Neanderthals shared a version of the *FOXP2* gene related to speech, and this is the best indication so far that Neanderthals could speak. [301]

More recent DNA analyses showed that *Homo sapiens* did indeed mate with Neanderthal. What's more, the Neanderthal-modern human mating apparently took place in the Middle East, shortly after modern humans had left Africa, not in Europe—as has long been suspected.

> "We can now say that, in all probability, there was gene flow from Neanderthals to modern humans," lead study author Ed Green of the University of California, Santa Cruz, said in a prepared statement.
>
> "That's no surprise to anthropologist Eric Trinkhaus, whose skeleton-based claims of Neanderthal-modern human interbreeding—previously contradicted with DNA evidence— appear to have been vindicated by the new gene study to be published tomorrow in the journal Science. "They've finally seen the light ... because it's been obvious to many us that this happened," said Trinkaus, of Washington University in St. Louis, Missouri, who wasn't part of the new study.'"[302]

[299]"DNA Sleuths," Ibid. p. 22A.

[300] A German study is underway to decipher the Neanderthal genome. Nicholas Wade "Scientists hope to unravel Neanderthal DNA and human mysteries." New York Times, July 21, 2006.

[301]Tina Hesman Saey, Team decodes Neanderthal DNA, Science News, March 14, 2009, p. 6.

[302] http://news.nationalgeographic.com/news/2010/05/100506-science-neanderthals-humans-mated-interbred-dna-gene/

Denisovans

Something important happened in 2010. A finger bone was discovered in a Siberian cave which defied categorization. With study, scientists determined this was a parallel human species, neither Neanderthal, nor *Homo sapiens*, but a species capable of mating with both:

> DNA evidence suggests they had dark skin, eyes, and hair, and had a Neanderthal-like build and facial features. However, they had larger molars which are reminiscent of Middle to Late Pleistocene archaic humans and australopithecines. Denisovans apparently interbred with modern humans, with about 3–5% of the DNA of Melanesians and Aboriginal Australians and around 7%-8% in Papuans deriving from Denisovans. Introgression into modern humans may have occurred as recently as 30,000 years ago in New Guinea, which, if correct, might indicate this population persisted as late as 14,500 years ago. There is also evidence of interbreeding with the local Neanderthal population, with about 17% of the Denisovan genome from Denisova Cave deriving from them. A first-generation hybrid nicknamed Denny was discovered with a Denisovan father and a Neanderthal mother. [303]

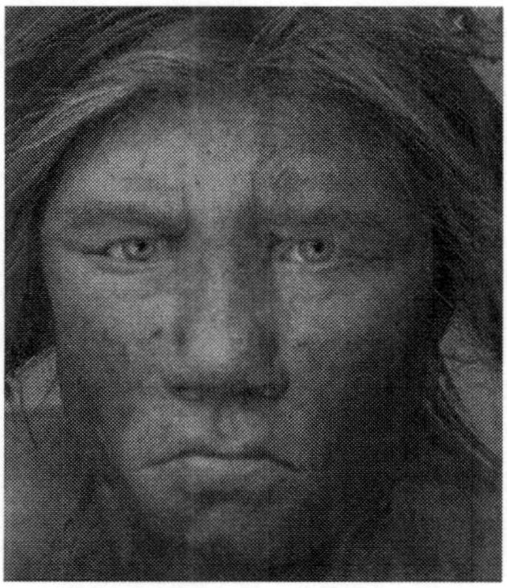

Denisovan. Source: https://www.nationalgeographic.com/science/2019/04/enigmatic-human-relative-outlived-neanderthals/

[303] Wikipedia Denisovan, https://en.wikipedia.org/wiki/Denisovan

How old are we?

With that short background we now take up one of our important questions, of just how old our species is. This is a second shock to our traditional understanding of evolution. First *Homo sapiens* differs from *Homo sapiens sapiens*, the former referred to as "archaic" and latter as "anatomically modern."[304] Cro-Magnon is a term used for the early or archaic *Homo sapiens*.

For the date for which fully modern *Homo sapiens* appears we go to the laboratories of the famed Steven Jay Gould. From this group, in an article entitled "Out of Africa vs. Multiregionalism," Ted Billings[305] reviews the literature searching for the oldest known *Homo sapiens sapiens*. Some skulls date between 92,000 and 115,000 years, while the Skhul fossils in Israel date from 80,000-119,000 years, and yet another from the same collection dates to 130,000 years.[306] [307]

Earlier textbooks placed anatomically modern man at only about 50,000 years of age, but the best evidence now pushes that date back to 130,000 years and counting.

Robert Sanders reports that "fossilized skulls of two adults and one child discovered in the Afar region of eastern Ethiopia have been dated at 160,000 years, making them the oldest known fossils of modern humans, or *Homo sapiens*." [308]

---

[304] Others suggest the age of Homo sapiens is 150,000 years. Lewin, Roger. New Scientist. 19 July 1997: 5.

[305] "Out of Africa vs. Multiregionalism" December 7, 1999 stevenjaygould.org

[306] Conroy, G. Reconstructing Human Origins: a Modern Synthesis. (W.W. Norton & Company: New York, New York & London, England) 1997 p. 439-440

[307] Leakey, R. "Early Homo sapiens remains from the Omo river region of south-west Ethiopia: Faunal remains from the Omo Valley" Nature 222(1969):1132-1133

[308] Robert Sanders, "160,000 year old fossilized skulls uncovered in Ethiopia are oldest anatomically modern humans, 11 Jone, 2003, Berkeley News, http://www.berkeley.edu/news/media/releases/2003/06/11_idaltu.shtml

But another skull, the Omo skull, discovered in 1967, was redated in 2005 at 196,000 years of age, also an anatomically modern *Homo sapiens*![309] The study was published in *Nature* in February 2005 using isotopic analysis of volcanic ash above and below the finds reports "our species, *Homo sapiens*, has a new pair of ultimate old-timers" and reports the age of both Omo I and II skulls at 196,000 years.[310] [311]

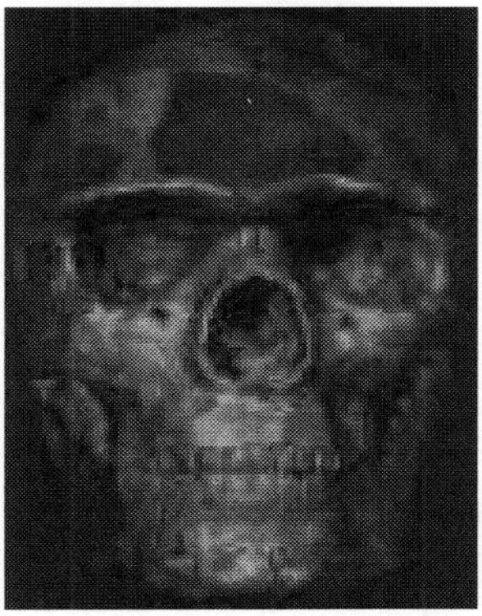

Omo I skull considered a modern H. sapiens and dated at 196,000 years

Add to that a recent cross-cultural DNA genome survey corroborates that *Homo sapiens* may indeed be 200,000 years old: Researchers found

> That the ancestors of the KhoiSan, hunter-gatherers living today in southern Africa, began to split off from other living humans about 200,000 years ago and were fully isolated by 100,000 years ago. That

[309] "Malcolm Ritter, "Ethiopian fossils dated to 195,000 years, called oldest remains of modern humans." Associated Press, The Miami Herald International Edition, Feb. 17, 2005, p. 10A. In addition " Fossils of early Homo sapiens previously discovered in 1967 in the Kibish region along the Ethiopian Omo River have now been redated to about 195,000 years ago by a team of American and Australian scientists including Ian McDougal, Francis Brown, and John Fleagle. According to the latest research results published by McDougall et al. (2005) in Nature, both skulls probably stem from chronologically comparable silty sapropel layers of the so-called Member I of the Kibish Formation. Their finding levels are now dated to about 195,000 years by a tuff layer positioned slightly below the fossils and providing a terminus post quem of 196,000 years on basis of its content of radioactive Argon." 2004 Athena Review, "Archeology in the News," https://www.athenapub.com/

[310] "Human fossils are oldest yet." Science News. Feb., 26, 2005, Vol. 167, p. 141.

[311] A recent piece in Science News may push our age as far back as 250,000 years. See Tina Hesman Saey, "First rough draft of Neanderthal genome released." Science News, March 14, 2009. Another unique study of mitochondrial DNA of the Sandawe people in Tanzania showed modern humans to be 170,000 years of age. Source: Ann Gibbons Science; 6/13/2003, Vol. 300 Issue 5626, p1641

finding hints that our ancestors already had evolved behaviors seen in living humans, such as language, 200,000 years ago.[312]

I wrote to one of the authors of that study, Prof. Joshua Akey, to confirm that there was no ambiguity on this controversial dating. Indeed, the answer came back that the latest estimate is the biologically, anatomically modern *Homo sapiens* has been here not 50,000 years, or 130,000, or 160,000 years, but 200,000 years! [313] A second study from Ethiopian fossils also supports dating anatomically modern humans to 200,000 years.[314]

That is far different reckoning than was taught in American high schools not that long ago.

Mungo Man, Chimps, and H. sapiens Genes

Another problem we want to look at concerns recently discovered Mungo Man. Let us get some background first on human DNA and chimps before we begin this discussion.

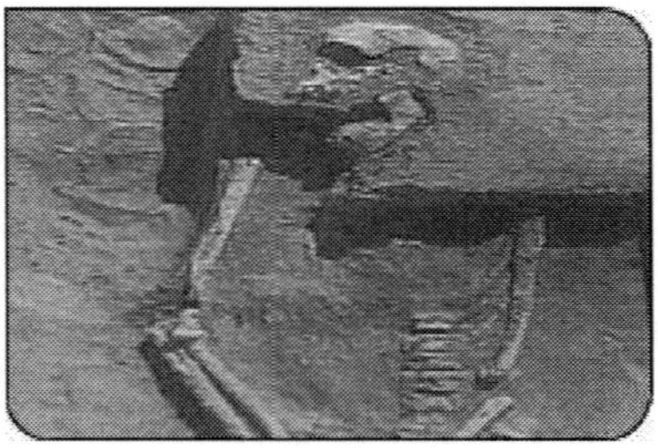

Differences in DNA between *Homo sapiens and* chimpanzees are approximately 0.6-1.4%. Some authors have speculated that of the 30,000 genes in biologically modern *Homo sapiens*, only about 300 (or 1%) differ significantly from chimpanzees.[315] [316] [317]

---

[312] "A single migration from Africa," New York Times, Sep 2016, http://www.nytimes.com/2016/09/22/science/ancient-dna-human-history.html?_r=0

[313] personal email.

[314] *Fleagle, John G.; Brown, Francis H.; McDougall, Ian (17 February 2005). "Stratigraphic placement and age of modern humans from Kibish, Ethiopia". Nature.*

[315] Will Hart, The Genesis Race, Rochester Vt: Bear & Co., 2003, p. 27; Note that since the human chimpanzee genome has not been deciphered, one wonders how this figure was estimated.

[316] It has been recently argued that some 1500 human genes appear to contain unique mutations compared to chimps. Source. Human genes take evolutionary turns, Jan. 17, 2004, Science News, p. 165.

[317] I shared the 'alien hybrid' hypothesis with a Harvard academician, Dr. Maryellen Ruvolo, a professor in Harvard's Anthropology Laboratory. Her comments, by no means validating the idea, were at least generous enough not to pooh-pooh it outright. Most interesting in our exchange of emails was that now that the human genome has been deciphered, scientists are rapidly on the way to deciphering the chimpanzee genome. Once this is accomplished, we should have a better idea of exactly which genes are distinctly human and divergent from chimp DNA. Presently we are only guessing at the level of differences between human and chimp but our research base is steadily improving. Since this email exchange decoding the chimpanzee DNA has been rapid and is described

A recent study looked at 97 specific genes to see what differences appeared between humans and chimps. This investigation revealed we have an even closer association with chimps sharing 99.4% of their DNA.[318] A more recent study using a different counting method, however, suggested that overall there might be as much as a 5% difference and thus only a 95% similarity between human and chimp DNA. [319]

The traditional argument, to which most of us were exposed, was that in the 4.8 million years separating chimps and humans, a gradual evolutionary creep pushed inexorably forward where incipient biologically modern man differentiated out greater cranial capacity and a more advanced neurology. All of that culminated in the emergence of cognitively superior, anatomically superior *Homo sapiens.*

Thus, the meager one-to-five percent genetic difference between chimps and man is the genomic *platform* upon which evolutionary biology explains how *Homo sapiens* emerged to take over and change the entire character of the planet.

Recently a discovery reported a 60,000 year-old *H. sapiens* skeleton, called Mungo Man. It is the oldest human skeleton ever to provide viable DNA. Morphologically a biologically modern man, its DNA was *substantially different from* our own, "something scientists had not expected and like nothing they have ever seen before." [320]

Mungo Man became the oldest human fossil that was physically similar to modern humans. But further analysis of Mungo's mitochondrial DNA (mtDNA) *could not link him to any human population living today.* [321]

The John Curtin School of Medical Research found that Mungo Man's skeleton's contained a small section of mitochondrial DNA. After analyzing the DNA, the school found that Mungo Man's DNA bore no similarity to the other ancient skeletons, modern Aborigines and modern Europeans. [322]

---

further in this chapter. [Note that Dr. Ruvolo is not the genetic research anthropologist cited later who reviewed the biological section and sought to remain anonymous.]

[318] Randolph E. Schmid "Study tightens chimp-human link" San Francisco Chronicle, May 20, 2003, p. A2.

[319] Britten, R.J. 2002. 'Divergence between samples of chimpanzee and human DNA sequences is 5% counting indels.' Proceedings National Academy Science 99:13633-13635.

[320] Leigh Dayton, N.Y.Times Syndicate, cited in The News, Mexico City, Jan. 15, 2001, p. 20.

[321] Fran Dorey, When did modern humans get to Australia, 18 April, 2024, Australian Museum https://australian.museum/learn/science/human-evolution/the-spread-of-people-to-australia/

[322] Brian Simpson,"What if the first humans came from Australia?" July 2024, http://www.convictcreations.com/aborigines/prehistory.htm

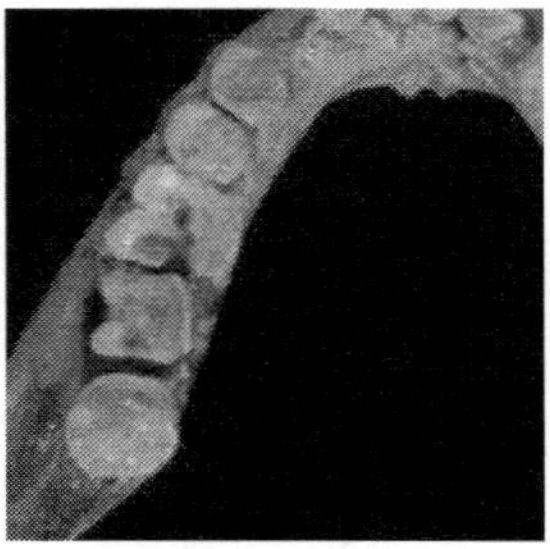

Another author said Mungo Man's of mitochondrial DNA has no modern match.[323]

Some argue that Mungo Man could not have migrated out of Africa and thus different lineages of *Homo sapiens* independently emerged (the multiregional hypothesis). A more recent study seems to suggest Mungo Man was outside the range of a human, namely that he may have represented a subspecies now extinct:

> Mungo Man [Willandra Lakes Hominid (WLH3)] and Kow Swamp 8 (KS8) yielded unique mtDNA sequences. Importantly, these two sequences fell outside the range of contemporary human variation.[324]

Other researchers believe that differing variants of *Homo sapiens* existed, some of which became extinct like Mungo Man. Ian Tattersall and Jeffrey Schwartz say fifteen different species of humans may have existed over time where Mungo Man would merely be one example. [325] Others stridently object to the nostrum of multiple and different sub-species[326] and say only a single group migrated out of Africa some 100,000 years ago [327]

Mungo Man is controversial. The evidence and DNA dating have been reviewed and the re-reviewed. The current opinion on this subject still holds that his DNA is far different from current *Homo sapiens* DNA, but the age of the specimen now has been re-dated at 42,000 years. [328] A more recent inves-

[323] Leigh Dayton. Mungo Man, ABC Science, 01 Jan, 2001 https://www.abc.net.au/science/articles/2001/01/01/2813404.htm

[324] Tim Heupink, et. al. "Ancient mtDNA sequences from the First Austalians revisted," PNAS06 June, 2016 https://www.pnas.org/content/113/25/6892

[325] Ian Tattersall & Jeffrey Schwartz, Extinct Humans. Westview Publishers, 2000.

[326] "Branchless evolution." Science News, April 15, 2006, p. 227.

[327] "Gene, fossil data back inverse human roots," Science News, Vol 159, Jan. 13, 20001, p. 21.

[328] Originally thought to be 62,000 years old . . . "The consensus was unanimous—Mungo Man was buried about 42,000 years ago, " says geologist Jim Bowler of the University of Melbourne who discovered Mungo Man in 1974 in the dry bed of Lake Mungo in New South Wales. Mungo Man is the oldest ritually buried skeleton in the world—his body was painted with ocher. Cited in "Finding Mungo Man." 14 Nov., 2017. The Guardian

tigation uses a different DNA analysis technique and says Mungo Man's DNA has been contaminated. [329] Others dispute that. [330] [331]

So, the DNA of modern *Homo sapiens* may have changed since the time of Mungo Man 42,000 years ago, or perhaps it did not. That controversy has not been settled. In other words, the significance of Mungo Man may be nothing at all if it's true that the specimen was contaminated. We don't know.

But it also may be that the *human genome itself has changed*, and that the earliest specimen of retrievable human DNA found in Mungo Man means the human genome *has significantly changed within the last 40,000 years.*

If we look at other studies of human DNA, a discovery of a tooth in Ethiopia dates to 8,500 years.[332] I have not found definitive evidence of earlier human DNA than Mungo Man. However, thanks to an evolutionary biologist who consulted with me, I learned of a Bulgarian sample of *Homo sapiens* DNA dating to 45,000 years.[333] A follow-up article showed the Bulgarian sample matches modern-day human DNA between 99.4% and 99.9%.[334]

On the other hand, a researcher I contacted at the Max Planck Institute for Evolutionary Anthropology told me in 2020 that he suspects the difference is greater than that, but the academic paper is still under peer-review and not yet released. [335]

So, whether it is Mungo Ban or Bacho Kiro, we can make one important conclusion: human DNA today is *different in some substantial ways from what it was 45,000 years ago.*

Since this might be getting technically a bit onerous, let us summarize our discussion up to this point: (1) *Homo sapiens* does appear related to Neanderthal, and was his contemporary up to about 30,000 years ago.[336] (2) Anatomi-

https://www.theguardian.com/australia-news/2017/nov/14/finding-mungo-man-the-moment-australias-story-suddenly-changed

[329] Dani Cooper, "New DNA technology confirms aboriginal peoples were the first Australians,"06, Jun, 2016 Science. http://www.abc.net.au/news/2016-06-07/dna-confirms-aboriginal-people-as-the-first-australians/7481360

[330] J. Lyon Laydens, "Science or Prehistory," 2017 https://prehistoricfantasy.blogspot.com/2016/09/why-mungo-mans-original-dna

[331] M.H. Monroe,"Australia, the land where time began," 23 Sep, 2011 http://austhrutime.com/genetic_evidence.htm

[332] Meagan Gannon, "Oldest human DNA from Africa reveals details about an ancient culture"27 Mar, 2018, Live Science. https://www.livescience.com/62131-oldest-human-dna-mysterious-ancient-culture.html

[333] Jean Jacques Heublin, et. al., Initial upper Paleolithic Homo Sapiensefrom Bacho Kiro Cave," 11 May, 2020, Nature, https://www.nature.com/articles/s41586-020-2259-z

[334]

[335] Personal email, name withheld.

[336] A study conducted by the Gibraltar Museum reports Neanderthals may have been alive in Spain as recently at 24,000 years ago. Malcolm Ritter, "Study: Neanderthals lived later than thought." Associated Press, September 14, 2006.

cally modern man has been here for 200,000 years. [3] The oldest anatomically modern *Homo sapiens* DNA ever retrieved, 42,000 year-old Mungo Man, appears to differ *substantially* in its mitochondrial DNA profile from modern humans, and even if Mungo Man's DNA profile has been contaminated, the Bulgarian sample shows that human DNA *has changed* from what it is today.

The evolution of human culture

Although there is some meager evidence to show Neanderthal may have used stone tools, the origin of complex human symbolic behavior is recent and drastic. If we date *Homo sapiens* cognitive superiority to empirically observable tangible behavior i.e., technology, written language, buildings, cities, and the first emergence of what we call culture and civilization, then our precocity is roughly 6,000-11,000 years old. That represents a *fraction* of the evolutionary time separating chimpanzees and Man (about 2 one-hundredths of one percent).

Here is a more practical glimpse of the issue. In this instance it concerns a stunning observation about Egyptian hieroglyphs:

> Remains from the pre-dynastic period around 3500 BC show no trace of writing. Soon after that date, quite suddenly and inexplicably, the hieroglyphs familiar from so many of the ruins of Ancient Egypt begin to appear in a complete and perfect state. Far from being mere pictures of objects or actions, this written language was complex and structured at the outset, with signs that represented sounds only and a detailed system of numerical symbols...and it is clear that an advanced cursive script was in common usage by the dawn of the First Dynasty....What is remarkable is that there are no traces of evolution from simple to sophisticated, and the same is true of mathematics, medicine, astronomy, and architecture...even the central content of such refined works as the *Book of the Dead* existed right at the start of the dynastic period."[337]

According to John Anthony West, an expert on the early dynastic period,

> How does a complex civilization spring full-blown into being? Look at a 1905 automobile and compare it to a modern one. This is no mistaking the process of development. But in Egypt there are no parallels. Everything is right there from the start.[338]

Where are the precursors to the hieroglyphs? And where are the documents spanning at least a couple of millennia corroborating a *gradual and*

[337] Graham Hancock, Ibid., https://www.goodreads.com/work/quotes/1709115-fingerprints-of-the-gods-a-quest-for-the-beginning-and-the-end

[338] Wikipedia "Egyptian pyramid construction techniques.: http://en.wikipedia.org/wiki/Egyptian_pyramid_construction_techniques

progressive development of hieroglyphs, pyramid building, mathematics, and astronomy?

If we are to explain all this using the traditional route, namely that some kind of genetic mutations happened which gave rise to our cognitive superiority, then we have to confront some gnarly issues. How did we get so smart *so quickly* if we have been around as long as we have?

If we share between 95-99% of our DNA with chimps, and the one to five percent differentiation [339] took 5 million years of evolution to accomplish, our differences in brain size, tool use, language development, and cognitive capacity should be explained somehow by these same genetic differences. If so, then we have to imagine rather dramatic mutations (a) *in an extremely small portion of our DNA* and *(b) in the very, very recent past.*

While you are doing the genomic gymnastics to think through that puzzle, consider what Carl Sagan says about the evolution of our big toe:

> The time scale for evolutionary or genetic change is very long. A characteristic period for the emergence of one advanced species from another is perhaps a hundred thousand years; and very often the differences in behavior between closely related species---say, lions and tigers---do not seem very great. An example of recent evolution of organ systems in humans is our toes. The big toe plays an important function in balance while walking; the other toes have much less obvious utility. They are clearly evolved from fingerlike appendages for grasping and swinging, like those of arboreal apes and monkeys. This evolution constitutes a respecialization—the adaptation of an organ system originally evolved for one function to another and quite different function—which required about ten million years to emerge. The feet of the mountain gorilla have undergone a similar although quite independent evolution.[340]

[339] A very recent study, once again corroborates that there is a difference of approximately 1.44 percent between chump and human DNA sequences. Science News, June 12, 1004, Vol 165, p. 382.

[340] Carl Sagan, The Dragons of Eden, New York: Random House, 1977 p.4

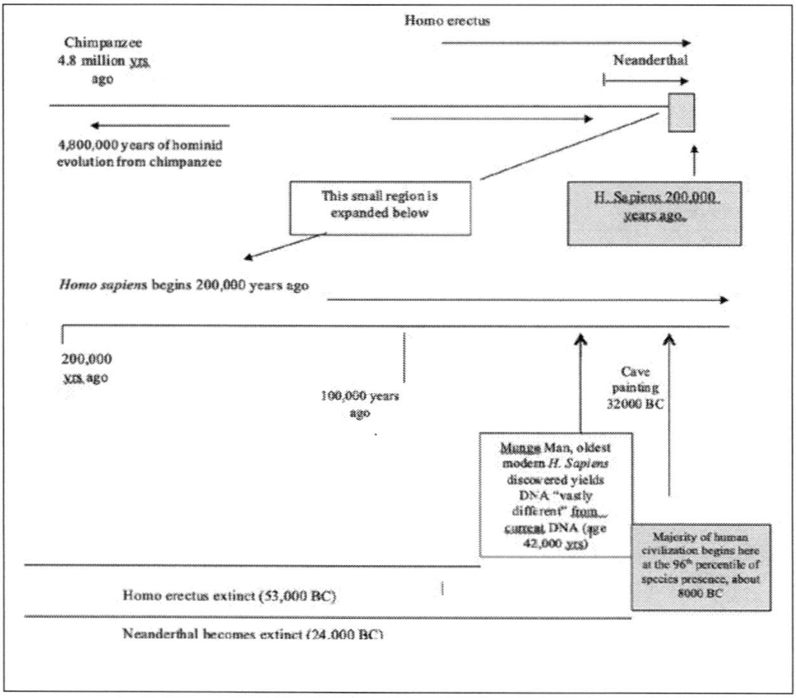

Fig. 1 Evolution of the chimpanzee and *Homo sapiens*. Graphic based, in part, on K. Wong, "African Exodus," *Scientific American*, Nov. 2000, pp. 82-83.

If it takes over a hundred thousand years to differentiate out a human big toe from a chimpanzee big toe, how can we use evolutionary biology to explain the *virtually instantaneous* emergence of human culture?

Of all the hominids, which diverged from chimpanzees, only one shows evidence of consummate intellectual gifts. The whole planet shivers and shakes as *H. sapiens* marches across its surface. From our first recorded civilization until the present, planet earth was altered, electrified, mined, industrialized, overpopulated, deforested, polluted, and warmed to the point of melting ice sheets at both poles *in less than ten millennia*, but our anatomically modern species has been here—at the latest count— for 200 millennia!

Something does not fit!

The hurdle over which evolutionary biology must jump is to explain this breakneck pace by a mysterious set of genetic mutations all occurring within that narrow band of DNA difference between ourselves and chimpanzees and in the narrowest chunk of Darwinian time possible.

Traditional explanations

The development of agriculture:

Let us first review what traditional anthropology says about this conundrum. In *Non-Zero, the logic of human destiny,*[341] Robert Wright explains mankind's dearth of culture for such a long period of his history. The essence of his argument is that as evolution manipulates genes to move toward greater complexity (single-celled to multicellular), cultural evolution uses 'memes' or ideas/innovations, which spread from the simple to complex in very similar fashion.

The spread is like a virus and disseminates through the human population steadily finally reaching exponential proportion. Thus human populations start small bereft of any major technological innovations, and then one meme starts to appear about every 20,000 years, followed by others every 1,400 years, until finally newer advancements happen every 200 years (like the invention of the comb). In this way Wright uses what he calls "ascending non-zero sum complexity" to explain the *hastening* of human culture, and by that he means through the exponential acceleration of cultural memes and innovations.

Moving from tribe to village, village to city, city to chiefdom, and chiefdom to empire represents another ascending series of memes which grow from simple to more complex. Wright places great emphasis on the development of agriculture spurring human innovation, beginning between 7,000 BC to 10,000 BC. This allowed more people to be freed from food gathering, created a larger, more sedentary and dense human population, and paved the way for even more innovations like writing, pottery, smelting, and science. [342]

One reason Wright's argument seems unconvincing is that it did not really take 200 years to develop a comb. It took over 190,000 years from biologically modern *Homo sapiens*' first appearance until the day he recognized he was having a bad hair day and decided to treat the condition.

Wright ignores the biological issue that if *Homo sapiens* was biologically intact 200,000 years ago—in terms of cortical development, for example—why did cultural memes of any genuine significance not really start appearing until the third and fourth millennium BC? Wright's book *explains away* the very abrupt beginnings of human culture by stretching them out and extrapolating backward in order to give cultural evolution a patina it doesn't

[341] Robert Wright. Non-zero: the logic of human destiny. New York: Random House, 2000.

[342] A very recent study asserts that evidence for agriculture; specifically the cultivation of wild grasses along with grains and cereals can be pushed back to about 23,000 years ago. See Seed s of agriculture move back in time, Science News, July 24, 2004, vol. 166, p. 61.

deserve. In this way the impetuous beginnings of human culture are made to seem gradual, progressive, and the result of ersatz uniformitarianism.[343]

The frontal lobes

Let's try another traditional approach:

This time a biological argument is used. *Homo sapiens*, it is said, has a significantly larger frontal cortex, and this represents the *sine qua non* for the human cultural big bang.

The expansion of the frontal lobes, compared to other primates, has been the mother lode from which most contemporary explanations flow. Supercharged, *Homo sapiens* frontal lobes historically provided the base for a geometric progression in conceptual, symbolic thinking, and *this single feature* differentiates us from lower primates including Neanderthal.

Unfortunately, this explanation has to confront some rather recent bitter facts. The *California Institute of Technology* reports:

> Researchers have traditionally theorized that the frontal cortex, a brain region linked to mental faculties such as planning and reasoning, expanded to an unprecedented extent during human evolution. However, a new analysis of brains from many different mammals takes the uniqueness out of our frontal cortex.
>
> Lemurs, gibbons, chimpanzees, and other primates have roughly the same proportion of brain tissue devoted to the frontal cortex as people do, say Eliot C. Bush and John M. Allman of the California Institute of Technology in Pasadena. Lions, hyenas, and other carnivores display a substantially smaller frontal cortex relative to the rest of the brain.
>
> ...Bush and Allman widened the scope of frontal cortex analysis, focusing on primates and carnivores. They compared 25 primate species with 15 carnivore species. Computerized analyses of a series of brain slices identified various neural regions and yielded volume estimates for them.

[343] Another problem in this argument is that in Giza, where the great pyramid was built, agriculture was not established in 10,000 BC, nor even by 7000 BC. There is no evidence of firmly established agriculture until 5000 BC, a thousand years after its appearance southwestern Asia, and quite late to be able to create a dense, highly populated region where the greatest edifice on earth would suddenly spring up. Source: Christine Hobson, The World of the Pharaohs, Ibid., p. 49. In addition Wright assumes memes like agriculture spread culturally and through learning, but recent evidence suggests that the spread of agriculture from early farmers in the Middle East into Europe was a function of gene transmission and intermarriage. "Farming spread into Europe via population mixing rather than by natives simply adopting agriculture." See "European face-off for early farmers." Science News, January 7, 2006, p. 14.

In all the primates, the frontal cortex displayed about the same relative size, approximately 36% of the total brain volume, while carnivores had less than 30% of their brain in this area.

Intriguingly, lemurs and other prosimians—regarded as the most primitive primate suborder—exhibited a slightly greater frontal cortex proportion than people and great apes did. That finding challenges the influential theory that the frontal cortex progressively expanded in our primate ancestors, Bush holds." [344]

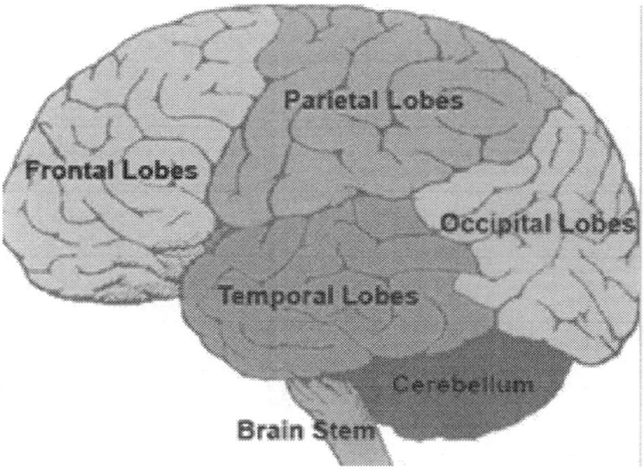

If the rug has been pulled out from under this traditional explanation of the human cultural big bang, a 2013 study drives the nail further into the coffin. Robert Barton, professor of Anthropology at Durham reports,

> Human frontal lobes are exactly the size expected for a non- human brain scaled up to human size...the size of human frontal lobes did not significantly increase relative to other primate species that human split off from millions of years ago.[345]

Traditional arguments indeed have problems, and many anthropologists admit to ardent head-scratching in trying to answer why human culture happened as quickly as it did after such a long period of dormancy. Trying to explain these rapid-fire events using traditional methods twists itself into a rhetorical pretzel.

---

[344] Bruce Bower, "Brain Size Surprise: All primates may share expanded frontal cortex." Science News, March 13, 2004; Vol. 165, No. 11 , p. 163

[345] "Frontal lobe size in brain does not explain human intelligence, 13 May, 2013 Medical Daily, http://www.medicaldaily.com/frontal-lobe-size-brain-does-not-explain-human-intelligence-245843

But mythology sits quietly off in the corner with its own rendition of these affairs. Let's take a closer look.

The mythological theory expressed

Let us recast the essence of the 'alien-hybrid' hypothesis we have been thinking about. It took 5 million years to quadruple the size of a chimpanzee brain so that it would fit neatly into a *Homo sapiens* skull. We have remarkable propinquity to our simian cousins,[346] but a distinct behavioral difference separates us from them. Consider what chimps do now, today, after 5 million years of evolution:

> Among things they've been seen doing: throwing rocks at leopards; attacking a fake leopard with sticks, taking twigs, stripping them of leaves, poking them down into a termite nest, then pulling them out and eating the termites; pounding nuts open with sticks and stones. Chimps even use sticks to brush each other's teeth. Some chimps crumple up leaves, turning them into sponges with which to extract precious water from the hollows of trees. Chimps also use leaves to wipe the last bits of delicious brain from the skulls of freshly killed baboons.[347]

Bon appetit!

But is this the entire panorama of their intellectual and cognitive repertoire? Not quite. Other studies report chimpanzees engage in twenty-six different practices, behaviors, and rituals which are passed down from one sub-group to the next, and these behaviors qualify as fitting the definition of "culture," i.e., a learned trait that is passed excsomatically (non-biologically). So with chimpanzee culture and the twenty-six behaviors that took five million years to develop and evolve, how do we reconcile *Homo sapiens* cultural evolution against this? [348]

Just as the skull of 200,000 year-old *Homo sapiens* is like our skull now, it is likely that so too was his neocortex. It is certainly logical to assume that when he first appeared right next to Neanderthal and *Homo erectus*, modern *Homo sapiens* had virtually the same brain size and cortical capacities that he has today. If true, then why was he so incredibly stupid?

[346] Randolph E. Schmid "Study tightens chimp-human link" San Francisco Chronicle, May 20, 2003, p. A2.

[347] Robert Wright, Ibid. p. 287.

[348] Recent studies of gorillas in northern Congo have also been found to use tools (a branch to test the depth of a river) to the surprise of investigators. See "Wild gorillas take time for tool use." Science News Oct 15, 2005, Vol. 168

Table 3 Late Bloomer

Invention/Advancement	Date first appeared	Years since beginning of Homo sapiens sapiens*
Domesticated dogs	16,000 ypp	182,000
First major city Golbekli Tepe	11,000 ybp	189,000
Domesticated, goats & sheep	10,500 ybp	188,500
Barley lentils, chickpeas cultivated	10,000 ybp	190,000
Domesticated chickens	11,500 ybp	190,500
Domesticated wheat	11,250 ybp	190,750
Domesticated pigs	9000 ybp	191,000
Domesticated horses	6000 ybp	194,000
Copper & bronze	6000 ybp	194,000
Cultivated potatoes	5400 ybp	194,600
Brewing beer	5500 ybp	194,500
First written language	5500 ybp	194,500
Stonehenge	5000 ybp	195,000
Invention of wheel	5000 ybp	195,000

*Using the age of biologically modern man at 200,000 years. Ybp= years before present. Source: The table is adapted, in part, from Jared Diamond's Guns, Germs & Steel, NY: Norton, 2005.[349] See also "Science News," Dec. 14, 2013, "Modern dogs originated in Europe"; Steph Yin, "Who first farmed potatoes," New York Times, Nov. 18, 2016; Carl Zimmer, "How the first farmers changed history," New York Times, Oct 18, 2016; and https://www.southampton.ac.uk/~cpd/history.html

If you look at the table, you can see how long it took for anatomically modern *Homo sapiens* to develop the technology needed to start human culture. It is baffling to the extreme why it took so long, and why our species was such a late bloomer.

Why did it take 195,000 years to think of inventing a wheel,[350] 190,750 years to think of growing wheat, or 188,000 years before he got the idea of domesticating animals which could till the earth and grow his crops?

Why did it take 194,500 years to create a written language or fashion copper and bronze?

Why did he twiddle his Paleolithic thumbs for so long?

[349] Sources for the domestication of horses come from Bruce Bower, "The Botai of central Asia milked marks more than 5,000 years ago." Science News, March 28, 2009.

[350]* Table based on dating anatomically modern Homo sapiens at 130,000 yrs.

From 200,000 years ago to 30,000 BC one really couldn't tell the behavioral or cognitive difference between Neanderthal and *Homo sapiens.* Both species used spears, harpoons, fishhooks, traps, hunted game in groups, used flaked stone tools, and neither displayed any evidence of having constructed significant living structures. Graham Hancock keeps looking, but he hasn't found anything. If one could find significant empirical evidence that one species was clearly superior or smarter than the other in this Paleolithic period, say from 50,000-30,000 BC, it would be a stretch.

From 20,000 BC, *Homo sapiens* shows an incipient sophistication in his use of tools and hunting techniques, but it is not until about 8-4,000 BC that we start to see a flowering of human culture and the beginnings of what we now call 'civilization'.

If we were clearly a far superior species intellectually, we might find *Homo sapiens* villages while Neanderthal languished in caves, but the record in Paleolithic times is muddled. In fact, in 30,000 BC Neanderthal and *Homo sapiens'* remains were found in the same cave in Saint Cezaire, France.

To say that one species was superior to the other *then* is questionable. Multiple numbers of anthropologists make that clear. [351] [352]

In other words, the evidence shows that *we were as uninventive as they were,* and by present standards, Paleolithic, biologically modern *Homo sapiens* and Neanderthals both appear to be dullards.

Modern *Homo sapiens* reached by 68,000 BC, Australia by 60,000 BC[353] and peripatetic anatomically modern man spread around most of the Old World, but hadn't yet codify a single written language...even by 3,500 BC!

> "We know that 20,000 years ago all people up to that time, approximately 100,000 generations---were hunter-gatherers living in simple egalitarian tribal units. By comparison, only 500 generations in total have depended on agriculture, and only 10 generations have lived in industrial civilization." [354]

That is a remarkably sudden growth spurt. In a flash, everything changed. Our imbecilic IQ seems to double overnight. Consider what one author writes about Sumeria:

> Before the Sumerians there were no bakers, harpists, carpenters, jewelers, artists, engineers, mathematicians, bureaucrats, or scribes.

[351] Parker, S. 1992. The dawn of man. Crescent Books.

[352] "Neanderthals take out their small blades." Science News May 13, 2006, p. 302.

[353] Strausbaugh L. and Sakelarisc, S. 2001 "DNA and Early Human History. Neanderthals and Early Humans: But did they mate?" Univ. of Connecticut. Presented at the Evolution Symposium, NABT Convention, Montreal 7 Nov. 2001.

[354] Will Hart, Ibid., p. 148.

All these innovations appeared for the first time in their cities between 3700 BC and 3000 BC. In addition to irrigation, kilns, and writing, they also invented the wheel, the chariot, bronze, sailboats, mathematics... astrology, schools, and the idea of trades and professions. In short, *they invented nearly the entire foundation of all future civilizations*---all in one fell swoop.[355]

Apart from Gobekli-Tepe and Jericho, prior to 4,000 BC *Homo sapiens* generally lived as nomadic hunters and gatherers, but in the blink of an eye, the Sumerian civilization popped up around 3700 BC, and in addition to all of the above, a second author rhapsodizes further: They developed—and these were all *firsts*—

> ...high rise buildings, streets, market places, granaries, wharves... temples, metallurgy, medicine, surgery, textile making, gourmet foods, the use of bricks....carts, ships and navigation; international trade; weights and measures; kingship, laws, courts, juries, writing and record keeping; music, musical notes, musical instruments, dance and acrobatics; domestic animals and zoos; warfare, artisanship, prostitution. [356]

One could make a very defensible assertion that the cognitive and intellectual achievements of *Homo sapiens* between 4,000-2,500 BC *were more significant than in all the prior 194 millennia combined!*

Think about that!

The author at the Great Pyramid

In such a short period of time our adrenalized species struts out its cognitive stuff out like it's a fourth-of-July parade. Some kind of uncanny metamorphosis converts this being into an awesome paragon of innovation. He/she builds cities, creates the highly complex and elaborate language of Egyptian hieroglyphics, boasts a staggering knowledge of astronomy, encrypts *pi* and *Phi* in the Pyramid of Giza, and

[355]Will Hart, Ibid., p. 74.

[356] Z. Stitchin, 1980 The Stairway to Heaven Avon Books, New York, p. 86.

manages in 2,600 BC—some say earlier some earlier—to engineer a pyramid that uses two million blocks of stone, enough to enclose St. Peter's, St. Paul's, and the cathedrals of Florence, Milan, and Westminster combined! [357]

> Over two million limestone blocks rise to the height of a forty-story building. Each baseline exceeds two and a half football fields. Standing on top, an archer cannot clear the base with an arrow.... Occupying an area of thirteen acres, the entire bedrock base has been carved to less than an inch out of level. It is oriented within a tiny fraction of a degree from the cardinal points. Outer casing stones and inner granite blocks fit with such precision that a razor blade cannot be inserted between them. Blocks weighing as much as seventy tons (about what a railroad locomotive weighs) have been lifted to the height of a ten-story building and mated to the next block with wondrous precision... All this comes from what was supposedly an agrarian society, forty-five hundred years ago."[358]

One of the very first structures built by man, it used stones hewn and moved by processes still under debate even in the 21st century. The building even appears to encode the perihelion of the sun. (Not bad for the same dufus who took 190,000 years to come up with a comb!)

Our ancestors do not seem to be *gradually* using tools, *gradually* developing language, or *gradually* using trial and error to build a few hundred flawed structures until they finally get it right.

It does not seem to happen this way.

In an article in *Science*, the respected paleoanthropologist Stanley Ambrose, aware of this anomaly, makes a pithy observation: "A mere 12,000 years separate the first bow and arrow from the International Space Station."[359]

Human culture bursts on stage and gives the impression of a qualitative leap forward more than any quantitative "creep" up some Jacobian ladder of development.

> And the pyramid of Giza, after standing in the sun for 5,000 years, enduring floods, earthquakes, continental drift, and decomposition, still boasts that its north points north and its south points south with

[357] Draghici Sava, "The great Pyramid of Giza" Medium, 2022, https://medium.com/@draghicisava/the-great-pyramid-of-giza-b55bc8837f8f

[358] Marshall Payn, "The case for advanced technology in the Great Pyramid," Forbidden History, Ibid., p. 270.

[359] Stanley Amborse, "Paleolithic technology and human evolution Science, 2 Mar, 2001. http://www.usatoday.com/news/science/aaas/2001-03-01-tooluse-evolution.htm

an *inaccuracy* of only 0.06%.[360] The pyramid which our idiot-savant builds will stand as tallest stone building on earth *for the next 43 centuries*, surpassed only by 18th century construction. And not to forget, our ancestor-eggheads built an almost identical set of structures on another pyramid plane far off in China.

Science writer Will Hart expresses the dilemma another way:

We are still fumbling around trying to understand the mysteries of the pyramid cultures of prehistory, and of how we made the quantum leap from the Stone Age to civilization in the first place! It does not add up....

As they had little to no experience with wild grains, how did they know what to do to process them, or even that they were indeed edible... How did Stone Age man suddenly acquire the skills to domesticate plants and animals and do it with a high degree of effectiveness? We find purebred dog species like Salukis and greyhounds in Egyptian and Sumerian art: How were they bred so quickly from wolves?

The Sumerians were making bread and beer five thousand years ago and yet their very close ancestor...knew nothing of these things and lived by picking plants and killing wild beasts. It is almost as if they were given a set of instructions by someone who had already developed these things.

....Circa 8000-5500 BC, the tribes in the Nile Valley were living in semi-subterranean oval houses roofed with mud and sticks. They made simple pottery and used stone axes and flint arrowheads. They were still semi-nomadic and moved seasonally from one camp to another. The vast majority of tribes around the globe were living in a similar state. How do we get from there to quarrying, dressing, and sixty-ton stones into the world's most massive structure, and in such a short time?

[360] The first exhaustive survey of the monument in modern times was carried out by Sir Flinders Petrie in 1880-2. He used the latest equipment of the time and approached his task with great thoroughness. He found that the sides of the pyramid were indeed lined up almost exactly with the cardinal points of the compass: north, south, east and west. The accuracy of this alignment is incredible, with an average discrepancy of only about three minutes of arc in any direction; this is a variation of less than 0.06 per cent. Gray Johnson Poole, "Sir Flinders Petrie,Britannica, 30 May, 2024. https://www.britannica.com/biography/Flinders-Petrie

> ...[W]here are the smaller-scale pyramids...much smaller? Where are the crude stone carvings that precede the sophisticated stelae? The slow evolution of forms, from simple to complex, is all that human beings knew, not mud and thatch-roof huts and then large scale architecture employing megalithic blocks of stone.
>
> ...[T]the developmental phases are simply not there. ...The problem is that there is no intermediate step between them and Sumeria and Egypt....The orthodox theories are starting to rely more on the "official pronouncements" of authorities rather than on well-argued and well-documented facts.
>
> We have reached a crisis in the field of anthropology, history and archeology, because the conventional theses are unable to solve an increasing large number of anomalies. The explanations are thin and threadbare and. . .unable to support their own weight. The pieces do not lock together and fit into a smooth coherent whole."[361]

Curiously both our Egyptian and Chinese pyramid builders *insist* they got their inspiration, their skills, and knowledge from extraterrestrial ancestors and forefathers, Osiris in the case of the Egyptians, and the 'Sons of Heaven' in the case of China. Sumerians attributed their civilization to visitations by the Gods Enlil, Enki, and their cohorts. And, as we documented in an earlier chapter, they are quite specific in saying that knowledge of medicine, surgery, mathematics, and the domestication of plants and animals came from the "Gods." [362]

In other words, it didn't come from Atlanteans, it came from extraterrestrial tutors (and ancestors).

But mainstream cultural anthropologists retort

> Oh no you didn't! We're still looking for an explanation, but we're quite sure you didn't get anything from the Gods. Your myths, epistles, cuneiform tablets, cylinder seals, and all those other recitals are just superstitious, deluded logorrhea. We'll get back to you when we've figured it out. Maybe it was agriculture and the recession of the ice age; maybe mutated frontal lobes, maybe non-zero sum meme acceleration, but we can tell you now, definitively and unambiguously, none of your achievements came from extraterrestrial sources. No gods were here, and none gave humanity anything, no writing, no commandments on tablets, no math, no science, no pyramid plans, no blueprints, no

[361] Will Hart, Ancient Agriculture: in search of the missing links. Forbidden History, Douglas Kenyon (ed)., Vermont: Bear & Co, 2009, p. 200-201

[362] Zacharia Sitchin, The 12th Planet Rochester, Vt: Bear and Co, 2002.

secrets of metallurgy...nothing...We discovered all of that, through trial and error and plodding gradualness. Sorry to disappoint, but your "gods" are just figments of your overactive imaginations!

If the explanation is to be found in the human genome, those genes took an awfully long time to express themselves after 195,000 years of dormancy. To reiterate: Neanderthal's cranial capacity is *larger* than ours, and he was here more than 200,000 years,[363] but never invented a wheel, any written language, any agriculture, or any kind of counting system by the time of his extinction. With our superior cognitive debut on this planet, we matched our retarded cousin's dullardry doltishly walking hand-in-hand for at least 190,000 years [364] before showing any genuine signs of brilliance and inventiveness.

So is it not problematic to anchor the florescence of *Homo sapiens* culture to a *gradual* evolution of the human genome?

The changes in our capacities are far too immediate to pull out that plum. We have to postulate a storm of mutations in an extremely short span of geological time that Darwin would barely have time to put on his raincoat.

The weird thing in all this discussion is that our ancestors *actually wrote to us about their transformation!*

They left us messages on walls and scribbled pictographs on rocks all across the planet. They buried papyri in caves. They gave us the *Pyramid Texts*, the *Nihongi*, Hesiod's *Theogony*, *The Epic of Gilgamesh*, the Indian *Rig Veda*, and *Mahabharata*, the Japanese *Kojiki*, the Mayan *Popol Vuh*, the Babylonian *Enuma Elis*, the *Turin Papyrus*,[365] the *Tibetan Book of the Dead*, the Korean *Legend of Dan Gun*, the *Aztec Song of the Dead*, Ethiopia's *Kebra Nagast*, the Persian *Shahnameh*[366] the Chinese *Shenxian Zhuan*, and the Sumerian cylinder seals. Indeed almost all of *Homo sapiens*' earliest writing dealt with the theme of humankind's relationship to celestial beings.[367] [368]

---

[363] Gore, R. & Blair J. 1989 Extinctions. National Geographic, June 1989.

[364] If one uses the 196,000 Omo skulls as the first date of anatomically modern Homo Sapiens, then the last 10,000 years of cultural advancement would occur after 186,000 years.

[365] The Turin Papyrus, now in the Egyptian Museum in Turin, states that the first rulers of Egypt were the gods themselves, followed by demigods and finally the people it called "the followers of Horus" . Christine Hobson, The World of the Pharaohs, London: The Paul Press,1987, p.22

[366] Andrew Collins. From the Ashes of Angels. New York: Bear & Co., 2001, p. 116. See also p. 35 for the Kebra Nagast.

[367] John S. Brownless, "Japanese historians and the national myths: the age of the gods," UBC Press, 2011

[368] "Creation myths of civilization," Leon's EFL Planet, 2007, http://www.bibliotecapleyades.net/ciencia/esp_ciencia_life23.htm

I'm not talking about creationism here, a naïve and crypto-conservative view of man's origin—and this book is not an attempt to rationalize any such accounting. To the contrary, *Homo sapiens* commentary on the subject of our extraterrestrial contacts is found in the oral and written records of *virtually every culture on this planet,* Egyptian, Greek, Japanese, Judeo-Christian, Mayan, and Chinese sources. From Zeus to Kronos, Dionysus, Krishna, Marduk, Xbalanque, Osiris, and Christ, the narrative of our ancestors coming from the heavens, or our being children of the Gods—or their apprentices—is ubiquitous.

Zeus mated with morals who bore him children—as many as 92. [369] Hercules was the illegitimate child of Zeus and a mortal woman, as were Perseus, Helen of Troy, and Minos.[370]

In Assyria Tukulti-Urta , a human was begotten by a god.[371] In Mongolia Bodonchar was born as a result of a union between the God Tengri and a mortal woman. [372] The gods Romulus and Remus who founded Rome were said to have been born of a mortal woman's celestial matings. [373] In Anatolia a god "mates with a human and produces offspring." The first Japanese emperor Jimmu was born of a star-god, Amerterasu. [374]In Greek Myth the goddess Aphrodite is impregnated by a mortal Anchises, and the child of that union named Apneas. [375] The Great Indian Epic Mahabharata depicts a god fornicating with a human and giving birth to Bhishma.[376]

On and on it goes.

[369] Vasiliki Moutzouri, 23 Aug, 2023, "Zeus, The Supreme God in the Greek Pantheon," Olympioi, http://www.greek-gods.info/greek-gods/zeus/stories/zeus-divine-mates/

[370] Roger Ebert, "When Greeg Gods breed with humans" RogerEbert.com, 12 Feb, 2012, http://www.rogerebert.com/letters/when-greek-gods-breed-with-humans

[371] Boslooper, Thomas, The Virgin Birth, The Westminster Press. 1962

[372] Gua Lococo Paul, Genghis Khan: History's Greatest Empire Builder, Potomac Books, 2008

[373] Richard Lewis A Deist's love of god and his natural world, Xilbris, 2015

[374] Kelly, Charles F. "Kofun Culture", Japanese Archaeology. 27 Apr., 2009.

[375] William Hansen, Classical mythology: a guide to the mythical world of the Greeks and Romans Oxford Univ Press, 2005

[376] "There are a lot of myths about male gods having sex with women," Quora, 2024 https://www.quora.com/There-are-a-lot-of-myths-about-male-gods-having-sex-with-women-Is-there-the-opposite-myth-goddesses-having-sex-with-male-humans

Australian rock art depicting intercourse

So while present scientific biases teach that we should treat this archaic literature as superstitious *Homo sapiens* graffiti, I have one politically incorrect suggestion: Those forefathers who putatively mated with us, or inserted a few snippets of DNA into our genome —or tutored us—and bequeathed moral codes, knowledge, written language, and cognitive precocity....maybe it is just possible they came from elsewhere.

Virtually all our mythology says that is exactly what happened. Hypothetically...tentatively...guardedly. . .cautiously...*why not at least think about taking them at their word?*

How terrible would it be if a handful—just a handful— of evolutionary biologists, archeologists, and anthropologists actually studied this hypothesis, legitimately *within* academia rather than always outside it? Why must it be so egregious to suggest that our apocalyptic rise out of 190,000 years of stone age-stupor into the brilliant beings that created the seven wonders of the world as rapidly as we did might, just might have been influenced by outside agitators?

Why should that hypothesis be treated as occult fantasy?

Perhaps because it is confused with evangelical creationism, and, if so, that is indeed unfortunate. Or perhaps the God of Denial in academia simply refuses to budge even an inch from its pedestal.

More on the efflorescence of human culture

Our other hominid cousins, Neanderthal and *Homo erectus* in particular, do not antedate us with gawky machines, wheels that cumbrously spin, or butterfingered settlements made out of the wrong kind of mortar. They just don't seem to possess these kinds of artifacts, edifices, languages, drawings, or tools. Neither did we for the longest time, and then suddenly we did. Bang!

Big Bang!

Specifically, Neanderthal wandered the northern reaches of our planet for 230 millennia, and we explored its surface for a good 190 with nary a hint of intellectual precocity. [377]

Neither of us.

That's a total combined period of mutual, shared, cognitive sluggishness of more than 400,000 years—and still neither came up with a bow and arrow, no one thought of domesticating a wild pig, nobody thought up a phonetic alphabet—even though recent genetic evidence suggests Neanderthal also was equipped for speech [378]—and neither species seemed smart enough to consider cultivating a potato.

If we were to ask what exactly *Homo sapiens* was doing which was superior to *Homo Erectus* or Neanderthal during all this time, we would be hard-pressed to list any real behavioral differences. A recent *Scientific American* article makes clear that it is entirely indeterminate if *Homo erectus*, Neanderthal, or *Homo sapiens* was smarter or more ingenious than the other.[379] Neanderthal apparently used fire, may have developed a flute, and used mammoth bones for building material, and he seems to have developed some stone tools and used claws and feathers for decoration, so he had *some* cultural sophistication.[380]

If brain size, tool use, language development, and cognitive capacity are going to be explained by evolutionary theory, it is necessary to postulate rather apocalyptic mutations in our DNA *suddenly*, and in the very, very recent past. Even 24,000 years ago, we find a cave which both Neanderthal

[377] John Darnton, Neanderthal, Open Road Media, 2014

[378] Marjorie LeMay, The language capability of Neaanderthal man," May, 1975, American Journal of Biological Anthropology,

http://onlinelibrary.wiley.com/doi/10.1002/ajpa.1330420103/full

[379] Ian Tattersall, "We are Not Alone" Scientific American, Dec. 1999, p. 56.

[380] Russell Tuttle, Neanderthal Culture, 11 Jun 2024, Britannica, https://www.britannica.com/topic/Neanderthal/Neanderthal-culture

and *Homo Sapiens* shared, and differentiating the two with respect to tool use or *any differential cognitive sophistication* is entirely indeterminate.[381]

Beginning only about 8000 BC [382] *finally* there are signs of animal and plant domestication and organized settlements, but it is really not until about 4,000 BC that human culture absolutely explodes. Beginning after the fourth millennium with the advent of the Sumerian and Egyptian civilizations to the present, a mere 6,000 year period, human cultural advancement exponentially accelerates.[383]

Physical anthropology insists that this human cultural eruption has to be explained by biology and genetics in some way. One way is to say that a mysterious set of rapid mutations gave us cognitive capacities much greater than Neanderthal, such as larger or more complex frontal lobes. Not only was our brain smaller than Neanderthal, but research cited earlier shows that there is no substantial or significant *qualitative divergence* in frontal lobe development in comparison to a number of other primate species.

A second rationale is based on the end of the ice age. The mainstream argument goes like this: With the recession of the ice-age comes farming. With farming, populations are no longer nomadic, and sedentary peoples create increasingly complex divisions of labor: populations increase, cities form, writing and technology develop. But if agriculture was the ultimate cause for the ascent in human culture, as radical environmental determinists suggest, [384] we are puzzled.

If you look at a map of the last ice age *at its maximum*,[385] there were plenty of areas still available to develop agriculture. Why didn't we? Dates, apples, and grapes were all growing, chimpanzees were eating something, but we weren't growing anything or cultivating anything during all this time.[386]

[381] Neanderthal debate goes south. Science News, Sept 23, 2006, p. 170.

[382] Jared Diamond, Guns, Germs & Steel, Ibid. p. 329.

[383] Recent investigations decoding Neanderthal DNA suggest they share 99.5% of their DNA with H.Sapiens, but investigators still do not believe any significant interbreeding occurred between such remarkably similar species. See "New DNA test yields Neanderthal clues, showing genome nearly identical to humans." New York Times, November 16, 2006, p. A10.

[384] See Jared Diamond's Guns Germs and Steel (Ibid) as an example.

[385] Courtesy of http://www.johnstonsarchive.net/spaceart/earthicemap.jpg

[386] Brian Palmer, "The real cave man diet," 12 Feb., 2012, Slate, https://slate.com/human-interest/2012/02/the-real-caveman-diet-what-did-people-eat-in-prehistoric-times.html

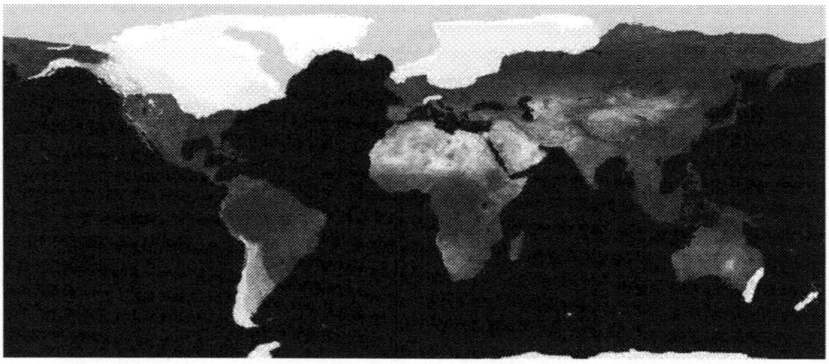

Ice age at its maximum

Why did it take agriculture so long to even start up in the first place (at least 190 millennia from our inception as a species).[387]

Secondly, if agriculture creates sedentary populations which, in turn, boost divisions of labor and intellectual achievement, why did great and flourishing agriculturally-based, sedentary civilizations, like the Inca Empire which stretched 2,500 miles across the continent of South America as late as 1200 AD, still never develop writing? [388]

The cultivation of maize has been pushed back to as early as 2,000 BC in Peru,[389] and yet even three thousand years later, no writing system appears. If the society was citified, sedentary, and unburdened from daily drudgery, where was the poetry and literature?

The Incas reached their zenith *forty centuries after* the Egyptians proudly mused over their hieroglyphs and the haughty Sumerians polished their cuneiform tablets, but Inca bookshelves were still empty. [390]

The step-by-step gradualness of cultural evolution seems to skip quite a few steps, if not trip and stumble along its explanatory path.

Connecting the dots

The flowering of human culture from its pyramids to this interconnected planet, begins—let's be honest—not much earlier than 8,000 BC. For

[387] See also Last Big Meltdown. By: Prendergast, Kate, History Today, 00182753, Aug2007, Vol. 57, Issue 8

[388] Indeed if agriculture precedes the development of writing, the development of agriculture in Peru dates to some 5,380 years ago, at the height of the Inca empire four millennia later writing had yet to appear. See "Gone with the flow," Science News, Nov. 12, 2005.

[389] Ancient Andean maize makers. Science News, March 2, 2006, Vol. 169, p. 132.

[390] Gilsdorf, T.E. (2000). Inca Mathematics. In: Selin, H. (eds) Mathematics Across Cultures. Science Across Cultures: The History of Non-Western Science, vol 2. Springer, Dordrecht. https://doi.org/10.1007/978-94-011-4301-1_11http://link.springer.com/chapter/10.1007/978-94-011-4301-1_11#page-1

a species that is at least 200,000 years old, the tardiness is staggering and suggests that something qualitatively striking occurred to alter us so much in so little time.

Beginning about the 94th percentile of *Homo sapiens* tenure on this planet, there was a quantum "great leap forward," from phlegmatic stone-age intellectual simpleton to an innovative, literate, pyramid-building genius...a metamorphosis that happened, frankly, in a Darwinian nanosecond.

If we date the vast majority of human cultural evolution as occurring in the last 12,000 years as paleoanthropologist, Stanley Ambrose [391] argues, and we use for our benchmark the latest research for the first appearance of anatomically modern *Homo sapiens* as 200,000 years, then our cultural emergence only truly *begins* in the 94th percentile of our time on this Earth.[392]

Prior to that we were as dimwitted as our simian and hominid relatives. Only in the most recent six percent of our time here has our anatomically modern species shown significant evidence of a gathering intellectual ability. This sudden ascent is nettlesome if it is to be explained by a gradual, developmental, evolutionary set of changes occurring *within* our species in this tiny sliver of time.

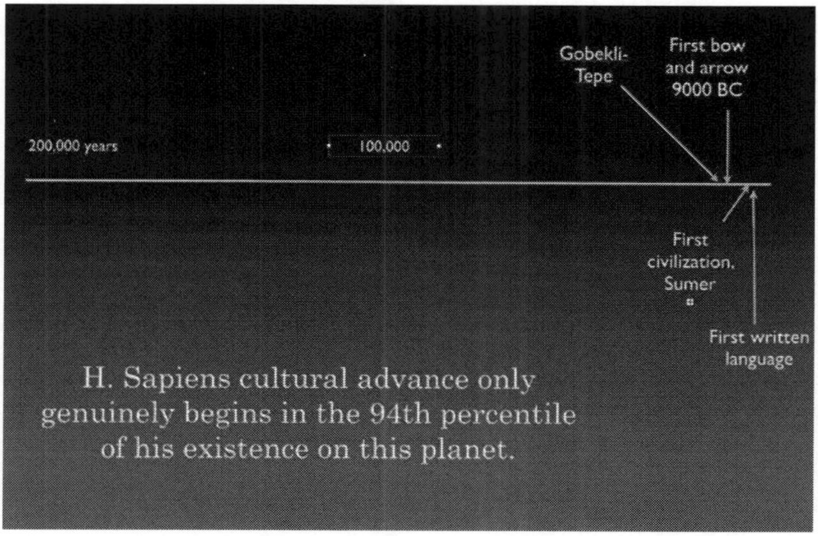

While modern anthropology at least dares to admit that it is perplexed by the furious ascent of human culture, it is curious that our mythological record is so blatantly ignored.

[391] Stanley Ambrose,Ibid.

[392] Wikipedia, Archery, https://en.wikipedia.org/wiki/History_of_archery

Physicists and astronomers don't ignore it. They have no difficulty consulting ancient books to fix the date and time of a supernova or to identify the dates of various earthquakes or droughts. In fact, one Sumerian cuneiform tablet accurately describes Halley's comet.[393] The physical sciences routinely use mythology as a reliable and accurate repository of human archaic memory, but when it comes to the issue of human origins, anthropology insists any such evidence is, of necessity, irrational, hallucinatory, and superstitious regardless of how prolific or cross-cultural.

A recent discovery in Babylon dating to 350 BC shows astronomers employed a form of precalculus in describing Jupiter's motion across the night sky.

> "Until now, credit for this kind of mathematical technique had gone to Europeans who lived 15 centuries later... This is a truly astonishing find," said professor Mathieu Ossendrijver, a professor at Humboldt University in Berlin... "It's a figure that describes a graph of velocity against time...that is a highly modern concept." [394]

Mythology is treated like a little elementary school child made to sit in the back of the room wearing a dunce's cap and admonished to "stay quiet." The message from our ancient literature has been rendered mute; the point of view it espouses is completely and thoroughly brushed aside.

But ignored, shunned, and sitting with a frown under that dunce's cap there is still a quiet voice murmuring. And that voice says that humankind has been influenced and changed as a result of visitations from outer space by Gods, beings, creatures, vessels, spirits, angels, and in sundry other entities from Ezekiel's space ships to the vessels that carried the pharaoh off with 33 other divine beings. Sure, there is a lot of diversity in these accountings, from one language to the next, one people to the next, and one epoch to the next, but the common subtext of these mythological narratives is that we have been designed, educated, altered, *and/or* are the actual offspring of extraterrestrial interaction.

[393] With respect to cutting a pink granite sarcophagus I contacted a mainstream archeologist and posed this question about diamonds not being available to Egyptian stone cutters, and the answer came back that they did it with stone tools, but no further explanation was given. On researching it four other substances could have done the job: beryl, topaz, chrysoberyl, corindum, or sapphire but it would have taken considerable time. An interesting text on Ancient Egyptian stone technology discusses stone saws Egyptians could have used and even has photos of errors in sawing stone with these tools.

[394] Kenneth Chang, "Signs of modern astronomy seen in ancient Babylon," New York Times, Jan. 28, 2016.

From Zeus to Kronos, Dionysus to Krishna, Osiris to Christ, the *Pyramid texts* to Hesiod's *Theogony*, the testimony that our species is the product of hybridization or tutelage from empyrean beings is ubiquitous.

But anthropology demands we ignore the misfit in the back of the room and premise everything man has done on evolution, DNA, and the human genome *without recourse to any outside intervention.* [395]

According to evolutionary geneticist Dean Hammer:

> The original humans' lifestyle was not terribly different from that of the Neanderthals. They were primarily hunters and gatherers. Then something changed. *Within an eye blink of geological time*, and for reasons that anthropologists still don't understand, man blossomed. Humans began to talk to one another. They began to paint and sculpt, to sing and dance. Soon they were farming and building permanent residences.[396]

The neurological complexity of dolphins is very close to *Homo sapiens* with respect to the ratio of brain size to body mass; its frontal lobes are similar—its cortex 40% larger—[397] yet in all the years that dolphins have been swimming around, close to 50 million years, we never discovered a dolphin settlement, city-state, underwater monument, religious temple, or any ongoing experiments in aqua-technology.[398] They do manage to pick up a sponge and use it as a tool in Australia to protect their sensitive snouts while foraging, and recent discoveries suggest that they talk to each other in sentences that could be as much as five words long, [399] but, so far, this is the only "culturally transmitted language or tool-use" ever observed from our neurologically sophisticated water-brothers.[400]

Beyond our inability to explain the breakneck rise of human culture premised on recent and massive still-unknown changes in the *Homo sapiens* genome, there are other problems to explain. We need to understand how the very first pyramids in Egypt were not of the stumbling, bumbling, one-of-these-days-were-going-to-get-it-right variety, but the first ones turn out

[395] A 1991 Gallup poll found that only 9 percent of the American public hold the belief that evolution proceeded strictly and solely through natural forces. See Dembski & Kushiner, Ibid, p. 43.

[396] Dean Hammer, The God Gene, New York: Doubleday, 2004, p. 202.

[397] Lori Martino, Comparison, of Primate and Cetacean Mentality 2011, Well Being International http://serendip.brynmawr.edu/bb/neuro/neuro98/202s98-paper3/Ball3.html

[398] Carl Sagan The Dragons of Eden, Ibid, p. 38.

[399] Krystal Johnson "Scientists record two daophins talk and listen to each other like humans, " 11 Sep, 2016, Yahoo News. https://au.news.yahoo.com/scientists-record-two-dolphins-apos-234321683.html

[400] "Sponge moms," Science News, June 11, 2005, Vol. 167.

to be some of the most incredible engineering achievements in the history of the human race, *far superior to any of the pyramids that followed!* [401]

Pyramid of Amenemhat I, built 6 centuries after the Great Pyramid (left)
Pyramid of Pepi II built 2 centuries after the Great Pyramid(rt)

It isn't supposed to work that way.

We don't start out with a Tesla and gradually evolve to the glorious Edsel.

But with pyramids that is what seems to happen. The most spectacular human construction, the Great Pyramid of Giza (Cheops) turns out to be superior to virtually all the pyramids built *afterward*. And these latter edifices were not improvements in structural engineering, but generally crumbling, ersatz, wanna-be imitations. Perhaps it is as mythology reminds us, *the Gods left!*

Inside the Great Pyramid is a sarcophagus made out of a solid piece of granite. Its measurements show an interior volume of 1166.4 and an exterior volume of 2332.8 liters, a ratio of external to internal volume that is *exactly* twice measured to the precision of a tenth of a liter. There is no indication that Egyptian craftsmen possessed any equipment which would have allowed this kind of precision. As for the cutting of the sarcophagus.

> [I}it must have been cut either with bronze saws studded with diamonds or saws of a much harder metal. But diamonds have never been found in Egypt, and at the same time the sarcophagus was supposedly made, metals harder than bronze were unknown to the Egyptians or anybody else on earth. [402]

Indeed one researcher took some modern tools, accurate to one one-thousandth of an inch, and tried them out on that sarcophagus.

[401] One pyramid, the pyramid of Dioser was built 80 years prior to the Great Pyramid.
[402] Whitley Streiber, The coming global superstorm. New York: PocketStar Books, 1999, p.61.

> Their smooth, flat surfaces, orthogonal perfection, and incredibly small inside corner radii that I have inspected with modern precision straightedges, squares, and radius gauges leave me in awe. Even though after contacting four precision granite manufacturers I could not find one who could replicate their perfection.[403]

I decided to fact-check these assertions and wrote to two granite manufacturers. I gave them the precise measurements of the sarcophagus and asked that it be cut from a single piece of granite. One said it would be extremely costly but that it could be done. I asked him how, with lasers, or with some kind of specialized cutting tools. He replied, "using diamond saws and diamond tipped tooling." Another manufacturer said, "at this point we cannot make your tub from one solid piece." So one says it can be done these days, another not; but note that diamonds were not discovered until the 4th century BC—in India. The question remains how the Egyptians managed to cut that pink granite sarcophagus 2,000 years before diamonds were first known. [404] [405] [406]

So when we apply an evolutionary perspective to all of this, as if our brains, capacities, technologies and culture *gradually and progressively* improve over time, our hard drives freeze up. With the Egyptians, nothing seems to happen the way it should.

Researcher and scholar Peter Tomkins writes about the Great Pyramid as follows:

> Till recently there was no proof that the inhabitants of Egypt of five thousand years ago were capable of the precise astronomical calculations and mathematical solutions required to locate, orient, and build the pyramid where it stands. It was attributed to chance that the foundations were almost perfectly oriented to true north, that its structure incorporated a value for *pi* accurate to several decimals and in several distinct and unmistakable ways; that is main chamber incorporated the "sacred" triangles. . .which were to make Pythagoras famous, and which Plato in his *Timaeus* claimed as the building blocks of the cosmos. Chance was said to be responsible for the fact that angles and slopes display an advanced understanding of trigonometric values, that its shape quite precisely incorporate s the fundamental proportion for the "Golden Section," known today by the Greek letter *phi*...According to

[403] Christopher Dunn, "Precision," in Forbidden History, Ibid., p. 255

[404] With respect to the diamond manufacturers contacted, they were Strargranite and Artistic Granite, Inc.

[405] "Mechanical Methods: Petrie's comments," Global education Project, no date, http://www.theglobaleducationproject.org/egypt/articles/petrie.php

[406] "History of Diamonds," Brilliance, no date, https://www.brilliance.com/education/diamonds/history

modern academicians, the first rough use of *pi* in Egypt was not till about 1700 BC—at least a millennium *after* the Pyramid; Pythagoras' theorem is attributed to the fifth century B.C.; and the development of trigonometry to Hipparchus in the second century before Christ... Recent studies of ancient Egyptian hieroglyphs...have established that an advanced science did flourish in the Middle East at least three thousand years before Christ, and that Pythagoras, Eratosthenes, Hipparchus and other Greeks reputed to have originated mathematics on this planet merely picked up fragments of an ancient science evolved by remote and unknown predecessors.

Whoever built the Great Pyramid, it is now quite clear, knew the precise circumference of the planet, and the length of the year to several decimals—data which were not rediscovered till the seventeenth century. Its architects may well have known the mean length of the earth's orbit round the sun, the specific density of the planet, the 26,000 year cycle of the equinoxes, the acceleration of gravity, and the speed of light.[407]

Not only did the culture that built the pyramids have the math and physics, incredible stone carving techniques, and an uncanny sense of the cardinal points of the earth in its construction, but their language, Egyptian hieroglyphs, appeared just as suddenly without any lengthy period of

[407] Peter Tompkins. Secrets of the Great Pyramid, New York: Galad Boos, 1971, pxiii-xiv.

proto-hieroglyphics preceding it and with few developmental precursors.[408] According to noted anthropologist Jared Diamond:

> Hieroglyphic writing appeared rather suddenly, in nearly full-blown form, around 3000 BC. Egypt lay only 800 miles west of Sumer, with which Egypt had trade contacts. I find it suspicious that no evidence of a gradual development of hieroglyphs has come down to us, even though Egypt's dry climate would have been favorable for preserving earlier experiments in writing."[409]

Beyond Jared Diamond, other scholars similarly conclude that Egyptian hieroglyphs and Sumerian cuneiform, which appeared just a century or two before, are entirely different from each other, with neither influencing nor evolving one from the other..

How curious is that from the beginning of anatomically modern *Homo sapiens*, roughly 200,000 years ago, 194 millennia pass before any written languages appear, and when they do, the first and earliest civilizations (Sumer and Egypt) display two entirely different writing systems even though the societies are only 800 miles apart.

[408] The earliest known writing comes from Uruk about 3300 BC in Sumerian clay tablets. Hieroglyphs appear about this time too, quite different, elaborated, highly aesthetic, and in the view of many scholars without antecedents. Source of photo and text: http://www.usu.edu/anthro/origins_of_writing/hieroglyph_aesthetics/

[409] Jared Diamond, Ibid, p. 232.

Earliest known writing. Sumerian cuneiform (left) develops at roughly the same time as Egyptian hieroglyphics, about 3000-3200 BC, the first known writing systems of humankind.

Make a mental note of the gaping problem in our mainstream logic: Modern anthropology would tell us that it takes 194 millennia of anatomically modern *Homo sapiens* dimwittedness before the species finally gets the idea of domesticating a horse—*that's 1,940 centuries*—but the time it takes from the very first appearance of Egyptian written language to the building of the Great Pyramid, which cuts, moves, and fits the largest blocks of stone on earth in ways that today we cannot even duplicate, *that* only takes only six centuries!

It is incomprehensible to reconcile that these structures and their monumental technology were built in a culture that between 400-600 years earlier *had no written language and, no symbolic mathematics.* [410] [411] In other words, only 4-600 years pass from the first appearance of hieroglyphic writing to the construction of the Great Pyramid, but it takes over 190,000 years from our beginnings as a species before we first get the idea of domesticating a goat.

[410] There is controversy about which language was the first written language (3400 BC, near Luxor, to 3200 BC, the first hieroglyphs). Bone and ivory tags, pottery vessels, and clay seal impressions bearing hieroglyphs unearthed at Abydos, 300 miles south of Cairo, have been dated to between 3400 and 3200 B.C., making them the oldest known examples of Egyptian writing. Source Larkin Mitchel, "Earliest Egyptian Glyphs", Archeology Vol 52, No. 2, March, 1999 There is also controversy about the actual date of the Great Pyramid, some say it was built in 2590BC others as early as 3200 BC. The figure cited above uses a consensus of when the Great Pyramid was built, 2850 BC, and a consensus of the first written language in 3400 BC.

[411] Richard Mattessich, "The oldest writing and Inventory tags of Ancient Egypt," Accounting Historians Journal, Jun 2002. http://www.jstor.org/stable/40698264?seq=1#page_scan_tab_contents

As TV cynic John Stossel, might say, “Gimme a break!”

Something strange happened to our species that seems *enormously* atypical and certainly not a gradual, developmental, progressive, unfolding of inventions, technologies, or memes.

A new twist is added in 2020…

To add more spice to this discussion, a recent report reveals a British engineer retrieved some wooden artifacts *from inside* the Great Pyramid in 1872. They are called “The Dixon Relics,” and they were found inside the Queen’s Chamber. They were recently carbon dated to 3341-3094 BC. [412]

The Dixon Relics

This would hypothetically move the actual date of the Great Pyramid’s construction back 600 years, and, if that were true, then we might think the Great Pyramid was built at the very same time that hieroglyphics first appeared! In other words, the building of the most magnificent structure on the planet and the very first written language to appear could have been contiguous events!

Curious new genes

Dr. Jonathan Pritchard from the University of Chicago reported recent results of a genetic study called “The Hap Project.” Researchers identified 700 regions of the human genome where genes appear to have been shaped

[412]Jessica Murray, “Lost artifact from the Pyramid of Giza found in a box in a cigar box in Aberdeen,” The Guardian, 15 Dec, 2020. https://www.theguardian.com/uk-news/2020/dec/16/lost-artefact-from-great-pyramid-of-giza-found-in-cigar-box-in-aberdeen

"within the last 5,000 to 15,000 years," and some of these genes appear related to cognitive functioning, [413] although the majority seems correlated with physical attributes like taste, smell, poison-detection, etc.

> The average point at which the selected genes started to become more common under the pressure of natural selection is 10,800 years ago in the African population and 6,600 years ago in the Asian and European populations.[414]

Reporting in *Science*, Bruce Lahn and his colleagues at the University of Chicago, identified two genes which have gone under substantial evolution in the last 37,000 years, one having to do with brain size, and an "ASPM variant" of that gene arising some 5,800 years ago and now found in roughly 30% of the species. [415]

A third line of recent research isolated the *HAR1* gene, differing from chimp DNA by only 2 out of 118 base pairs, and this also seems to be a recently evolved gene that could be related to reasoning and learning.[416]

Another study reminds us that the genomes "of any two individuals on the planet are more than 99.5% the same." Despite that similarity, researchers have noticed that seven percent of human genes "fit the profile of a recent adaption with most of the change happening from 40,000 years to the present." In this study roughly 100 may be associated with cognitive functioning i.e., neurotransmitter genes including serotonin, glutamate, and dopamine *"with the majority emerging in just the past 10,000 years."* [417]

Many notable specialists, among them Francis Crick, assert that we should be cautious in interpreting this data.

Another study from the *Proceedings of the National Academy of Sciences* speaks of the "extraordinarily rapid recent genetic evolution of our species" which it links with the growth in human culture. [418] One new and very recent variant in our genome originated about 5,800 years ago and now...

> shows up in approximately 30% of today's humans. These time windows are extraordinarily short in evolutionary terms...Each variant

[413] Nicholas Wade "Still evolving, human genes tell new story." New York Times, March 7, 2006, p. A1.

[414] Nicholas Wade, "Still evolving Human Genes tell new Story, New York Ties, o7 Mar, 2006. http://www.nytimes.com/2006/03/07/science/07evolve.html?_r=1

[415] Genes tied to recent brain evolution. Science News, Sept 24, 2005, Vol. 168, p. 206. Note that critics hold these new alleles do no confer any new cognitive advantages.

[416] "Evolutions DNA differences." Science News, August 19, 2006, p. 117.

[417] "Human evolution kicks into high gear," 10 Feb. 2009, Science News. http://www.nbcnews.com/id/29123062/ns/technology_and_science-science/t/human-evolution-kicks-high-gear/

[418] John Hawks, Eri Wang, et. al., "Recent acceleration of human adaptive evolution," http://www.pnas.org/content/104/52/20753.short

> emerged around the same time as the advent of "cultural" behaviors. The microcephalin variant appears along with the emergence of such traits as art and music, religious practices, and sophisticated tool-making techniques--which date back to about 50,000 years ago. The ASPM variant coincides with the oldest-known civilization, Mesopotamia, which dates back to 7000 BC.[419]

In fairness, *none* of these authors attribute our evolving genome to anything extraterrestrial, but they do note that something very peculiar happened to the human genome in very recent times. Explaining our delayed cultural advance will probably take on the character of proposing these kinds of recent genetic alterations like *HAR1* or *ASPM*.[420] Together these discoveries could represent the 'smoking gun' for *Homo sapiens* exponential cultural emergence, the springboard from which human culture jumps after its long period of dawdling.

> The Social Science Genome Association Consortium, an international collaboration involving dozens of university labs, has identified a handful of regions of human DNA that affect cognitive ability. They have shown that a handful of single-nucleotide polymorphisms in human DNA are statistically correlated with intelligence, even after correction for multiple testing of 1 million independent DNA regions, in a sample of over 100,000 individuals.[421]

An even more recent study reports there are 538 genes associated with cognitive capacity and 187 regions of the genome associated with thinking skills. [422]

Certainly, it is possible these modifications evinced themselves in a manner entirely governed by the process of natural selection, but mutagenic changes usually take eons more evolutionary time. Recall that it took roughly 1.5 million years for the *Homo erectus* cranium of about 1000 cc to evolve into our 1350 cc skull.[423] To have hundreds of modifications in the human genome—so very recently—raises an eyebrow.

[419] "University of Chicago researchers find human brain still evolving, 7 Sept, 2004, UChicago Medicine.http://www.uchospitals.edu/news/2005/20050908-humanbrain.html

[420] Unfortunately a June, 2006 report suggests ASPM does not exert an influence on brain size. "Evolving genes may not size up brain." Science News, June 3, 2006, Vol 169.p. 349.

[421] Stephen Hsu, "Super intelligent Humans are coming," Nautilus, 2 Oct, 2014 Humanhttp://nautil.us/issue/18/genius/super_intelligent-humans-are-coming

[422] "Found, more than 500 genes are linked to human intelligence, 12 Mar, 2018, New Scientist. "https://www.newscientist.com/article/2163484-found-more-than-500-genes-that-are-linked-to-intelligence/

[423] Leakey, R. and Lewin, R., 1992. Origins Reconsidered, Abacus Books, London, pp. 58-64.

Agreed, professor Bruce Lahn may not yet genuflect at the altar of alien genetic intervention, but he does at least admit the human brain is the result of "extraordinarily fast evolution."[424]

Certainly, these variants could have arisen from a mysterious, unprecedented, turbocharged evolutionary process entirely governed by natural selection, but they also might have come from extraterrestrial manipulation as well. Could it be possible they occurred as a result of outside interaction or intercourse with the beings our mythologies have rhapsodized?

Has the human genome been tweaked, or *edited*, with extraterrestrial snippets here and there?

Already our species, that is human beings, in 2019 experimented with editing the genetic makeup of a rat and actually succeeded in improving its cognitive abilities. [425] Who is to say a similar kind of experimentation might have happened only this time we were the lab rats?

The ancient Egyptians told us where the hieroglyphs came from. They didn't come from Atlantis. They didn't come from trial and error. They didn't come from a millennium of gradually improving doodles. They came from the gods, from the sky. They were *imported*, the hieroglyphs, that is the words *mdju netjer* mean "words of the gods."

Ditto for the Chinese who identified the source and builders of their White Pyramid complex not from any mariners of Atlantis, nor from a millennia of piling up dirt with little buckets until someone stood up and said "Gosh, we've got a pyramid here!" No, they credit their gigantic mound and its celestial alignment to their *ancestors from the sky*. [426]

And they footnote the development of agriculture and medicinal herbs to the "divine farmer," Shennong, another sky god. [427]

But it is *de rigueur* in anthropology to dismiss all of that testimony as nothing more than occult fantasy.

Bragging rights

On their elaborate tablets and steles, the Sumerians might have told us of their long, enduring, and frustrating developmental struggles to create that advanced system of weights and measures, how they painstakingly cultivated botanical domesticates generation after generation, or how they spent a few centuries perfecting the very first musical instruments.

---

[424] Alok Jha, Human brain result of extraordinarily fast evolution, 28 Dec 2004, Evolution,https://www.theguardian.com/science/2004/dec/29/evolution.science

[425] Ellen Neff, "Crispr improproves prospects for transgenic rats," Nature Briefing, 21 May, 2019 https://www.nature.com/articles/s41684-019-0316-8

[426] Wikipedia. Chinese Gods, https://en.wikipedia.org/wiki/Chinese_gods_and_immortals

[427] Wikipedia Shennong, http://en.wikipedia.org/wiki/Shennong

Just as we inscribe our elementary school books with the names of Edison, Eli Whitney, and Henry Ford, how proper if the Sumerians and Egyptians gave credit where credit was due. How helpful if they serenaded us and celebrated the names of all their great inventors and developers, in metallurgy, shipbuilding, musical transcription, and ceramics, *but they didn't.*

None of these cultures wrote of themselves in this manner.

The Egyptians tell us explicitly that their knowledge came from the stars. Why should we reflexively throw out that testimony without a shred of doubt that they might have been trying to tell us something?

> From far Antiquity the Egyptians believed that Osiris and Isis descended from the skies to bring a wondrous civilization to the Lands of the Nile...The Book of the Dead vividly described the Shining Ones from the stars...after great cataclysms the Gods returned to those glittering constellations bejeweling the night leaving their descendants, the God-King, to govern Egypt."[428]

The Sumerians referenced their gods Enlil, Enki, and other sky gods for their knowledge and advances in agriculture,

> Without Enlil, the great mountain, no cities would be built, no settlements founded. No stalls would be built, no sheepfold established. No king would be raised, no high priest born.[429] [430]

The goddess Inanna (circa 4,000 BC) is presented the arts of civilization by the God Enlil:

> "O name of my power, O name of my power,
> To the bright Inanna, my daughter, I shall present . . .
> The arts of woodworking, metalworking, writing, toolmaking,
> leatherworking....building, basket weaving." [431]

Inanna graciously received those gifts, and there are no alternate cuneiform tablets that celebrated that *actual, mortal* inventors of woodworking, smelting, and writing, but all advances were invariably credited to the gods.

[428] W. Raymond Drake, Ibid., p. 183.

[429] Kramer, S. N. History begins at Sumer. Philadelphia: University of Pennsylvania Press, 1989, p. 92.

[430] Curiously recent work on the spread of agriculture out of this area reports that there were striking anatomical differences between the advanced farmers who migrated out of the Middle East and their European hunter-gatherer cousins. See "Cultivating Revolutions," Science News, Feb. 5, 2005, Vol. 167.

[431] "Myths of origins," Internet Sacred Text Archive, http://www.sacred-texts.com/ane/sum/sum07.htm

By the way, just as the Egyptians list of pharaohs begins with the gods followed by demi-gods, Sumeria also boasts a list of kings, and, as you can probably guess, the list also begins with sky-gods.[432]

So from all these accountings—Greek, Sumerian, Chinese, Mesoamerican, Egyptian, Japanese, Persian, Indian—we see ubiquitous mythologies trying to tell us something: *Homo Sapiens* appears to be the heir of a few centuries or millennia of tutelage, and may indeed be a creature whose genome was manipulated, amplified, edited, or influenced in some manner through intercourse with these same visitors.

Ashur, one of the Sumerian's pantheon of sky Gods

Back to psychoanalysis

Jung differed from Freud especially with respect to dream interpretation. Freud was preoccupied with finding out how the psyche *hides* things from itself, and thus the manifest content of the dream was a distorted representation of a deeper, latent, *hidden* truth. Jung interpreted dreams too, but he also suggested that sometimes a dream tells you exactly what it says; all you need to do is to look at it clearly.

Joseph Campbell, the mythologist, was influenced by Jung and believed myths could be interpreted or seen the same way we look at dreams. To him, a dream is a personal myth and a myth a collective dream. Consider, for example, the mythology of the Nativity, Christmas, and the manger.

About 31% of all *Homo sapiens* alive today—some 2.2 billion people—celebrate Christmas.[433] Nativity scenes are constructed across the planet and a

[432] Kingship and the Gods: A Study of Ancient Near Eastern Religion as the Integration of Society and Nature . Oriental Institute Essay. Chicago University of Chicago Press, 1948

[433] Bill Chappell, World Muslim population will surpass Christian, Pew Poll says," 2 Apr,2015, NPR.http://www.npr.org/sections/thetwo-way/2015/04/02/397042004/

very large chunk of the human race memorializes this event. So, taking on a Jungian perspective, what exactly are we looking at, honoring, or memorializing when we stare at it?

What do you actually see here?

We see three Wise Men wearing mitres or crowns who followed a star which directed them to a blessed and holy scene, a birth. But it was not just an ordinary birth. It was the birth of a *hybrid child* whose mother was an earthling and whose father a celestial being, a heavenly father.

Is the star Sirius? Are the three Magi, the three stars of the Orion Belt? Perhaps, but that is where "interpretation" begins.

What is *not* subject to interpretation is that humanity is celebrating the birth of a human being whose heavenly father—whose paternity and ancestry—comes from the stars.

That birth, St. Paul the Apostle, says ushers in a "new man" and that is not singular, but meant to imply a "new humanity." [434] To Paul, we are all "children of the Gods." And religious literature reminds us the meaning stretches beyond just one birth and one child:

"You are My Son: today I have begotten you." (Psalm 2:7)

"You are all sons of God" (Galatians 3:28)

"God created Man in his own image" (Genesis 1:27)

muslim-population-will-surpass-christians-this-century-pew-says

[434] Systematic Theology, Volume 2 by Wolfhart Pannenberg 2004

"You are all sons of the Most High" (Psalm 82:6)"

"You, O Lord, are our Father." (Isaiah 63:16)

"[Ye are] the sons of the living God." (Hosea 1:10)

"Our Father which art in heaven." (John 20:17)

"We are the offspring of God." (Galatians 3:26)

"We are the sons of Allah and His beloved ones." (Koran Al-Maeda Verse No. 18)

Today, any tenure-track anthropology professor who thought of taking that hypothesis seriously, namely that the *Homo sapiens* genome reflects interaction with an extraterrestrial source—the gods—either through interbreeding or some kind of extraterrestrial gene editing, would be gingerly escorted out of the ivory tower in a heartbeat.

Simply not going to happen.

The symbolism of the Sphinx & creation myths

Oedipus is asked to solve the riddle of the Sphinx. "What walks on four legs in the morning, two legs at noon, and three legs in the evening?" He answers "Man!" Man crawls on all fours in infancy, walks upright on two legs in adulthood, and uses a cane in old age. Riddle solved.

The riddle of the Sphinx, so attractive to Freud and the ancient Greeks, is solved by the assertion that *the Sphinx is a structure that represents Man.*

While Oedipus was looking at a different sphinx entirely in 350 BC, Jung alerts us to interpret symbols by just looking at them and just asking *what we see.*

The Sphinx that experts currently date to about the same time as the Great Pyramid, is, to this day, the most maximalist sculpture on earth over thirteen stories tall and a city-block long. And what does it depict?

It shows the head of a man or god attached to the body of a beast, claws and all. Perhaps that is exactly what the ancients wanted to memorialize. The "beast" here is one *Homo sapiens* animal indistinguishable from Neanderthal in intellectual dexterity. He had something attached to him, however, a head, a divine nature, an astonishing intellectual capacity that metamorphosed and transformed a hunter-gatherer half-wit into the most awesome reigning species on earth.

The Sphinx is a hybrid. Jesus himself is a hybrid, half man, half god. Just as so many ancient Greek heroes were the offspring of the Immortals, is the riddle of the sphinx saying the same thing about humankind?

Archetypes, genes, and hard-wiring

If *Homo sapiens* is some kind of alien hybrid, do we have any more proof for such a speculation? Actually no, not directly, but Harvard geneticist Dean Hammer says he found something peculiar: *Homo sapiens* spiritual sense, his tendency to look towards the stars and the Gods, might actually be genetically based.

For a psychological trait to be considered genetically rooted, it must pass through a rigorous methodological filter. Take homosexuality as an example. When data is obtained on the frequency of adult homosexuality in pairs of identical twins and then *compared to the frequency of that trait in fraternal twins*, whenever a statistically significant *discrepancy* occurs, the most reasonable explanation for the divergence falls at the feet of the genetic invariance shared by identical twins, but not fraternal.

Using this methodology, homosexuality, for example, has been shown to possess a 'heritability' factor of 0.50, meaning its occurrence can be explained as roughly 50% nature and 50% nurture. "Obesity" and "criminality" also passed through the identical-fraternal twin filter, and both are considered partially genetically based, albeit possessing lower heritability scores.

Using the procedure with the reliably measured, psychometric trait of "spirituality," Hammer finds its heritability almost as high as homosexuality or 0.43. That is to say the spiritual sense in our species appears to be an *inherited predisposition.*

> The correlation for identical twins was .37, about double the correlation of .20 for fraternal twins. When these numbers were analyzed, the estimate for heritability came out to 43%. In other words, nearly half of the reason the twins felt religion helped them, spent time privately praying, and had a sense of God's presence, was inherited. Since these twins were raised by different parents, in different neighborhoods, and sometimes even in different religions, their similarities seemed to be the result of their DNA rather than their environment.[435]

If spirituality is a genetically based, inherited predisposition, then one might easily pose the following dilemma: if religious belief is delusional and absurdly magical, how did it get into the genome?

Ever-skeptical evolutionary biologist, Richard Dawkins, refers to religion as "a virus of the mind...(from) a parasitic group of myths and falsehoods that serve no biological function or advantage." [436] If *Homo sapiens'* propensity toward religious belief is really the expression of a maladaptive and aberrant genetic mutation, what adaptive purpose could such a mutation have served all this time?

Is it possible the Gods not only have been delivering aphorisms, parables, caveats, and moral codes *verbally* on tablets and scrolls from umpteen mountain tops, but also etch-a-sketching on the human genome and grafting a divine imprint in a way not possessed by other creatures on earth?

Curiously, even in the Bible there is a cryptic reference to such hard-wiring in which God has inscribed his code and "written in their hearts." [Romans 2:13-15] John Calvin called it *sensus divinitatis*, a "sense of deity

[435] Hammer's study is based on a large sample of some 1383 subjects. Hammer has even suggested that VMAT2 gene is the specific gene in question. Quotation is from Hammer, Ibid., p. 144.

[436] Hammer, Ibid, p. 146.

inscribed in the hearts of all.[437]" Herodotus said the same in 350 BC. Ditto for virtually all primitive cultures, which seem to possess religious mythologies—and *cognitive imprints*—pointing to the stars. In Ecclesiastes, it is written "God has set eternity in the hearts of men, yet they cannot fathom what God has done from beginning to end."

Human mythology has a long list of sky gods who interacted with man, but we could add into this mix all the ancient creation myths that, in one variant or another, describe how creatures from the heavens created man. Not all cultures possess such myths that look up to the heavens, *but the majority do.* [438] This table lists quite a few more.

Ainu	Kagba
Algonquin	Krachi
Altair	Lanoan
Arikan	Lugbara
Assyrian	Malozi
Bagob	Mande
Boshon	Maori
Buriat	Minyong
Celtic	Modoc
Cheyenne	Mosetene
Dhammai	Nandi
Dogon	Negritos
Dyah	Ngombe
Efik	Ngurunderi
Ekoi	Nyamwezi
Fang	Pawnee
Fulaini	Popo
Hopi	Skagit
Hottento	Society Isle
Huron	Swahili
Ijaw	Tahilian
Inca	Thonga
Iroquois	Wakaranga
Joshua Indians	Wapangwa
	Yoruba

Table 4. Cultures with creation myths pointing to the heavens.

[437] Cited in Ronald Nash. Faith and Reason, MI: Zondervan Publishing House, 198, p39 See also William Dembski and James Kushiner, Signs of Intelleigence, Grand Rapids, MI: Brazos Press, 2004, p. 53.

[438] David Leening, A Dictionary of Creation Myths, New York: Oxford University Press, 1994. See also: The Origin of Life and Death: African Creation Myths: Ulli Beier (ed). London: Heinemann, 1966.

There are two ways to think about this discussion. One is to consider the *Homo sapiens* archetypal religious fixation as an aberrant genetic mutation that has caused our species to adhere to a delusional and magical set of maladaptive beliefs which are mysteriously fully resistant to the corrective surgery of natural selection for eons and eons or, to the contrary—that the Gods *left a genetic footprint in the form of an archetype.*

As one writer put it: "When tribes living in remote areas come up with a concept of God as readily as nations living shoulder to shoulder, it is fairly strong indication that the idea is preloaded in the genome than picked up on the fly....[439]

Carl Jung articulated the concept more eloquently He said about God—or the gods—they have. . .

> "...made an inconceivably sublime and mysteriously contradictory image...without the help of man, and implanted in man's unconscious as an archetype...not in order that theologians of all times and places should be at one another's throats, but in order that the unpresumptuous man might glimpse an image, in the stillness of his soul, that is akin to him and is wrought of his own psychic substance. This image contains everything he will ever imagine concerning his gods or concerning the ground of his psyche." [440]

[439] Jeffrey Kluger, "Is God in our genes," Time, 17 Oct, 2004, http://www2.gol.com/users/fmyers/Media/Courses/IH2/Natural%20History%20of%20Peace/5_God%20Gene_Time).pdf

[440] Martin Buber On Psychology and psychotherapy, Syracuse Univ. Press, 1999, p.64

Conclusion and Synthesis

Our birth is but a sleep and a forgetting
The Soul that rises with us, our Life's Star
Hath had elsewhere its setting
And cometh from afar
—William Wordsworth

Certainly, one quintessential idea in Part III of this text is that mankind's sense the gods and another world beyond this can be understood as a *memory trace*, preserved in myth, of contact with extraterrestrial civilizations which interacted with or even mated with our ancestors.

We have become far more intelligent than our simian brethren, not because we have a few more neurons than *Homo Erectus* or because we have up to a five percent edge in DNA ahead of chimps, or even because agriculture developed allowing humanity to flourish *finally* after an insipid 190 millennia of hunting and food-gathering.

Something *weird* happened in *very recent history* that is behind the cultural paroxysm since 12,000 BC.

That *Homo sapiens* is an alien hybrid, or perhaps the heir of an apprenticeship of extraterrestrial pedagogy, is a *competing hypothesis* that purports to account for our tempestuous cultural big bang.

Our origins are not merely of this world. They stretch back in time to touch ancestors who were not merely our chimpanzee brothers and sisters but may implicate a progenitor *not of this planet.*

Perhaps mankind's need to feel important or wanted or free of death, pain, and misery, prompts those divine aspirations, and maybe that is why our mythologies were concocted in the first place. But is it not curious that these heavenly notions and our star-studded bias seem to exist everywhere and for all ages past and present? More interestingly, we cannot find any such ideas in very similar creatures. Neanderthal, in 2,300 centuries of nomadic wandering, never created a cathedral much less an altar.[441] There are no *Homo erectus* sacrifices[442] to the gods of which we can be sure, and certainly no dolphin, bonobo, or orangutan behaviors betray any concern for a divinity or the heavens.

> Jane Goodall records how the death of an adult male chimp, who broke his neck falling out of a tree, led others to display intense excitement and anxiety around the corpse, even throwing stones at it...yet as close as chimps and other higher-order primates are to us biologically, they appear not to indulge in any elaborated behavior towards the corpse which is simply discarded and left. [443]

Only our species, it seems, drags out the incense and the red ochre to paint its face with untold designs and tattoos to adorn a conspicuous symbolism that solemnizes a thousand after-death rituals, pointing to the stars, pointing to our ancestors from the sky, and pointing to life elsewhere. What makes matters worse, it manages to carry on these dances, sacraments, ceremonies, and spectacles virtually *everywhere on earth* and has engaged in such practices as far back in time as we can extrapolate. Today some 84% of the human race believe in some form of divinity.[444]

The cult of the Caroline Islands built a memorial, a straw airplane and a runway which was a replica of the plane that brought John Frum to them.

[441] There is some debate concerning whether Neanderthal had a religious sense. In one gravesite, circa 80,000 BC, Neanderthal skeletons were found neatly placed in fetal positions decorated with flowers, jewelry and tools, indicating perhaps a sense in an afterlife journey. Other Neanderthal gravesites, however, show scattered and gnawed bones suggesting Neanderthal practiced cannibalism and was casual with the bones of his brethren. Despite this anthropological debate any significant or momentous religious artifacts simply have not been found. Source: "New Theory challenges Neanderthal 'flower burial' theory in shanidr cave," 3 Sep, 2023,Archeology News, http://faculty.rpcs.org/reillyl/Neanderthal_handout.htm. Another study reports that of 200 Neanderthal skeletons, only 30 showed some rudimentary signs that mortuary or burial practices might be in evidence. Source: Mike Pearson, The Archeology of Death and Burial, Ibid.

[442] Although there have been no religious artifacts associated with homo Erectus, some anthropologists argue that Erectus did make engravings in bone, perhaps decorative, and that some symbolic life might be argued to belong to this early homo species. Alexander Marshack, "On Paleolithic Ochre and the Early Uses of Color and Symbol," Current Anthropology, 22:2, April 1981, p. 188

[443] Mike Pearson. The Archeology of Death and Burial. Texas A & M PuPress, 1999, p. 148.

[444] "84 percent of world population has faith," 23 Dec, 2012 Washington Times, http://www.washingtontimes.com/blog/watercooler/2012/dec/23/84-percent-world-population-has-faith-third-are-ch/

Their rituals recapitulate and memorialize that event as they await the promise of their sky-god to return with more precious "cargo."

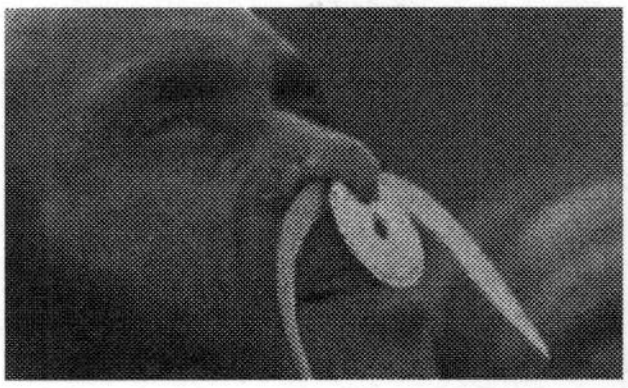

John Frum worshippers look to the heavens. Above they have constructed a straw replica of his as they await the return their god's return.

Humankind does the same thing—and far earlier —constructing strange and unusual monuments all across this planet, tracing lines large enough to be visible deep in outer space, and worshipping gods who come from stars and star systems which are identified *as the source of his ancestral origin, as well as his destiny.*[445]

All this happens while his cousins, from chimps to Neanderthal, all with coextensive brain structures, almost identical DNA, and often transplantable organs, keep their noses to the ground, preen themselves, forage for food, and languish in their natural habitats blissfully intact without any acknowledgment of deities, divinities, or the vaguest superstitious interest in the stars.

[445] Although a large number of ancient mythologies from all continents have creation myths which point to an extraterrestrial intervention or beginning, there are many which do not and which treat man's origin as coming from a cosmic egg, spirits from the underworld, the Raven, the Serpent, etc. See Creation Myths.

Our rituals and myths often are ways of remembering, of memorializing important events in our collective history, whether it is a planetary flood—which myriad cultures record in their respective mythologies—or visitations from people of the sky.

Homo sapiens did some marvelously suspicious things as testament to his belief in his forefathers coming from elsewhere. Consider again the Nazca lines of Peru. We know that these lines when seen from the air (photo to the right) depict creatures, birds, and are so large they can only be seen by flying above them in a plane. Recently, however, NASA released photos of other Nazca lines, significantly larger, which can be seen best not from airplanes but from outer space. These perfectly straight lines (left) are roughly 4 miles by 7 miles in length.

Academic political correctness

When you search academic databases for 'human extraterrestrial origins,' you find no articles at all. I do not know of any academics presently studying the issue of whether the *Homo sapiens* genome may have had an extraterrestrial influence. Such papers are entirely nonexistent in anthropology and archeology databases.

On Google, however, if one punches in "alien hybrid," there are 38 million hits, and if you use "mankind extraterrestrial origin" you get 3 million, so there is considerable *popular* interest in these topics, but in both the anthropology and archeological databases, there are *fewer than ten articles*, and all of these deal *solely* with geological evidence for extraterrestrial meteors.

We only need recall the problems of Nikolaus Copernicus, Johannes Kepler, Galileo—and even Sigmund Freud—to remind ourselves that simply putting a hypothesis on the table can cost a person their reputation. And in some cases more than that. After putting Giordano Bruno to death for suggesting the earth revolved around the sun, it took the Vatican 400 years before they got around to absolving him of his crime. [446]

In these days of political correctness, we have to admit science is shackled and bound to an agenda that places many ideas, like the one we are pursuing in this text, *totally* outside its margins.

Personal sidebar

In the case of this book, one valiant research microbiologist and genetic anthropologist kindly reviewed the biological section of this chapter for accuracy. I paid her a fee for an independent assessment of this chapter. I wanted to make sure my characterizations of biological data were not inaccurate nor my references too outside the margins. She answered quite emphatically, "No!"

That was reassuring.

When I asked her how the extraterrestrial hypothesis might be experimentally tested. She said, "If extraterrestrial contact had a mutagenic influence, ancient DNA samples would look significantly different from DNA sampled today."

This idea was certainly underscored in the Mungo Man discussion where mitochondrial DNA was surprisingly different 42,000 years ago than it is today; ditto for the Bulgarian DNA sample. 447 I then asked her whether I could cite her as a source,

[446] W. Raymond Drake, Ibid., p. 26.

[447] Since she reviewed this chapter, another DNA study (2016) of Pacific Islanders found whole sections of DNA that did not match any currently known human species, Stephen Morgan, "DNA of unknown human species found in Pacific Islanders," 25 Oct 2016, Digital Journal http://www.digitaljournal.com/tech-and-science/science/proof-of-unknown-human-species-found-in-dna-of-pacific-islanders/article/478035

> but she deferred and said I should simply refer to her generically as a "research genetic anthropologist."
>
> I asked "why?"
>
> She said she had a reputation to protect.

So my "anonymous" external reviewer privately thought the hypothesis was reasonable and even had empirically testable edges, but for reasons of her career and professional standing, she remains nameless.

This is the ethos in which this idea is tendered. The hypothesis that species *Homo sapiens* might have been schooled or that the human genome might have been genetically altered by extraterrestrial contact regrettably *is found nowhere in serious academic discussion.*

In sum, while we have looked at two major cataclysmic events beginning in 12,000 BC, events that caused the extinction of myriad species and almost took humanity down with it, we also discovered, unlike Graham Hancock's life-long quest, that there wasn't any significant antediluvian advanced civilization that existed before these traumatic events happened.

Maybe one day we will find Atlantis and revise that statement, but as it stands, whether Gobekli-Tepe, Gunung Padang, or the Yonagami monument, it is still hypothetical to suggest some gigantic advanced human culture was the precursor to the magnificent advancements made by humans later on.

Those advancements, from the Great Pyramid, to the beginning of written language, happened in the blink of a geological eye, and the *cause* of this unusual and rapid efflorescence in human culture was not the melting ice, not the emergence of agriculture, not the Atlanteans, but just as our mythologies have been trying to tell us all along, *the arrival of the gods.*

Forbidden, debunked, debased, and the butt of parlor jokes everywhere, this, nonetheless, is the most likely explanation for the origin of human innovation, human culture, creativity, and consciousness.

Our progenitors were extraterrestrial.

We are children of the gods.

Our father, who art in heaven, hallowed by thy name.

About the Author

Jerry Kroth, Ph.D. is an Associate Professor Emeritus from the Graduate counseling psychology program at Santa Clara University. He taught psychotherapy and personality theory, dreamwork, and research methods. Dr. Kroth has an abiding therapeutic interest in working with dreams, personal oracles, and the applications of dream theory to psychohistory. Jerry has been a member of the International Psychohistorical Association since 1983.

Dr. Kroth's seventeen prior books were in the areas of counseling psychology, child sexual abuse, learning disorders, metapsychology, research methodology, the Kennedy assassination, the Lindbergh kidnapping, UFOs, and collective psychology. In addition, he has written and presented over 75 papers on anxiety, child development, mass psychology, synchronicity, experimental studies of the dream process, the psychology of propaganda, psychohistory, and collective psychology. His most recent book is *The Psychic Immune System: a hidden epiphenomenon of the body's own defense.* Professor Kroth lives in California with his wife and two daughters and maintains a website: collectivepsych.com.

Picture Credits

All photos are public domain (PD) unless otherwise indicated. Those not footnoted in the text are listed here.

p. 2 Woolly Mammoth, Wikipedia

p. 2 Interglacial graphic: Uncertainty in empirical climate sensitivity estimated 1850-2017, PD, Jamal Munshi, Sonoma State University, page. 7.

p. 3 Giant Sloth: https://www.newscientist.com/article/human-arrivals-wiped-caribbeans-giant-ground-sloths/

p. 4 Extinct animals: https://en.wikipedia.org/wiki/Late_Pleistocene_extinctions

p. 11 Younger Dryas Graph source Research Gate https://www.researchgate.net/figure/Climatic-change-over-Earths-postglacial-period-Last-Glacial-Maximum-LGM-using-data_fig4_366988982/download?_tp=eyJjb250ZXh0Ijp7ImZpcnN0UGFnZSI6InB1YmxpY2F0aW9uIiwicGFnZSI6Il9kaXJlY3QifX0

p .13 Comet striking, aftershock: https://www.youtube.com/watch?v=zni7-0cSGDs

p. 13 The Black Mat Source. Live science, https://www.livescience.com/comet-impact-neolithic-period-human-civilization.html

p. 14 Nanodiamond, PD https://en.wikipedia.org/wik

p. 17 Willamette Meteorite: https://en.wikipedia.org/wiki/Willamette_ Meteorite#/media/File:At_American_Museum_of_Natural_History_2024_005.jpg

p. 19 Supernova Wikipedia

p. 22 fossilized tree rings https://www.iflscience.com/ancient-tree-rings-reveal-the-most-powerful-solar-storm-weve-ever-found-71042 (Cecile Miramont)

p. 23 Miyake solar storm

p. 28 Younger Dryas, two events PD

p. 29 Woolly Rhino, Wikipedia PD

p. 32 Clovis people stone tools Wikipedia PD

p 36 Nebraska bone bed PD: https://paleo-tourist.typepad.com/.a/6a0134898e4732970c01b8d2e112b2970c-pi

p. 36 Texas bone bed: PD https://www.texasbeyondhistory.net/bonfire/images/p1.html

p. 36 Ukraine bone bed: https://www.texasbeyondhistory.net/bonfire/images/p1.html

p. 38 Isaac Newton, Wikipedia PD

p. 41 Joseph Campbell, Wikipedia PD

p. 44 Sir Arthur Evans, Wikipedia, PD

p. 45 Atlantis, PD https://upload.wikimedia.org/wikipedia/commons/thumb/1/16/The_landscape_of_Atlantis.jpg/1024px-The_landscape_of_Atlantis.jpg

p. 49 Right angled corner stones Egypt & Peru PD https://www.facebook.com/p/Ancient-Buildings-Megalithic-Monuments-100063568813785/

p. 50 map drawing of Atlantis, https://esotericorigins.substack.com/p/atlantis-and-osiris-part-1

p. 53 Atlantic Ocean during Ice age https://ifunny.co/picture/map-of-the-atlantic-ocean-during-the-ice-age-when-Y5yK9Cns9

p. 53 Bimini Road, PD Wikipedia

p. 55 Distance from Atlantis to Peru and Egypt PD

p. 55 Photo of Atlantic Grid, PD, NOAA

p. 56 Lost city of Cuba PD https://historyenhanced.com/cuban-underwater-stone-structures-challenge-conventional-archaeology/

p. 57 photo submerged structure Cuba, PD personal correspondence

p. 58 Underwater land bridges between Yucatan and Cuba for 20,000 yrs, PD. https://www.researchgate.net/profile/David-Anderson-44/publication/263924190/figure/fig5/AS:667158030336609@1536207261564/The-Caribbean-and-Gulf-of-Mexico-ca-20-000-cal-yr-BP-showing-how-movement-throughout.png

p. 60 Oronteus Fineas maps (Kroth, Algora, 2001)

p. 63 Oronteus Fineas map with Antarctica Kroth, Ibid.

p. 63 Tributaries of Oronteus Fineas map. Kroth Ibid.

p. 63 Satellite view of frozen Ross Sea PD https://x.com/PreetChandi10/status/1741484657630237017

p. 66-67 Gobekli Tepe photos (2): https://theothertour.com/gobeklitepe/

p. 69 Natufian map Wikipedia PD

p. 70 Yonaguni photos: https://ru.pinterest.com/pin/ 848787861004347095/

p. 71 Yonaguni photos PD Ibid

p. 71 Carved head, Yonaguni, p. 61 Carved head, Yonaguni, https://www.facebook.com/1stWoUAnAH/photos/below-is-an-image-of-a-head-found-on-the-yonaguni-seabed-of-japan-within-the-sam/401857585954281/?paipv=0&eav=AfYdqpb74gyxGovaQy8XIFJ_QHi0BGQalVGgQYpzScrb6AW-NOWeFBqSGe4zzI6T2BHc&_rdr

p. 72-73 Sandstone tablet, from Dr. Prieto with permission

p. 75 Gunung Padang fantasy pyramid: https://ahotcupofjoe.net/2023/08/gunung-padang-what-archaeology-really-says/

p. 77 Gunung Padang pyramidal structure and alleged chamber

https://ahotcupofjoe.net/2023/08/gunung-padang-what-archaeology-really-says/

p. 61: Atlantis: Atlantis: https://www.youtube.com/watch?app=desktop&v=touhTJNoc8w&embeds_referring_euri=https%3A%2F%2Fmiski-liusuria.blogspot.com%2F&feature=emb_imp_woyt

p. 63 right-angled stones: https://www.pinterest.com/pin/856035841644862081/

p. 63 right-angled stone: https://www.pinterest.com/pin/858358010209665963/

p. 65 Gulag Padang: https://www.southeastasianarchaeology.com/2023/08/29/unlocking-the-real-mysteries-of-gunung-padang-separating-fact-from-fiction/

p. 68 photo of Gobelki-Tepe stone tools: PD https://bikeclassical.blogspot.com/2014/11/gobekli-tepe-beginning-of-history.htm

p. 83 Roswell book, Kroth

p. 85 Nazca Lines https://aneclecticeccentric.wordpress.com/2013/05/27/lines-of-nazca-peru/

Chapter 16: the photos in this chapter (pp 116-191) also appear in Kroth, *Aliens and Man*, New York: Algora 2002, pp 15-45. Sources: p. 57 courtesy of www.eyakimono.net and www.japanesepottery.com; p. 58. Satellite image by GeoEye.; pp. 63, 64; info@adventuresunlimitedpress.com; p.66. Library of Congress; p. 69 Courtesy of Eric Broder; p. 72 (a) (b) (c) copyright world-mysteries.com; p.76 courtesy of info@adventuresunlimitedpress.com; p. 79 Josef F. Blumrich; p. 89 source: http://www.ufocasebook.com/bestufopictures5.html

p. 86 Nazca drawing, The Astronaut: source Wikipedia; https://en.wikipedia.org/wiki/Nazca_lines#/media/File:L%C3%ADneas_de_Nazca,_Nazca,_Per%C3%BA,_2015-07-29,_DD_46.JPG

p. 86 Japan statue with goggles. https://www.japancuisine.net/japan-history-the-origins/

p. 87 Sycamore gliders: https://www.youtube.com/watch?v=FArWVgmamCE

p. 87 Egyptian helicopter source: https://www.tiktok.com/@mystery.magazin/video/6987117979271957765

p. 88 Pakal, Wikipedia PD

p. 89 UFO hieroglyph https://www.instagram.com/reel/C7009xquOy3/ https://studiovni.com/existe-t-il-des-references-historiques-aux-ovni-dans-la-bible-dans-dautres-textes-anciens-ou-dans-des-peintures/

p. 89 UFO hieroglyph https://www.quora.com/Do-any-prehistoric-cave-drawings-or-paintings-depict-spaceships-aliens-or-other-futuristic-things

p. 89 UFO hieroglyph https://www.pinterest.com/pin/152840981074805738/visual-search/?x=16&y=16&w=468&h=177&cropSource=6&surfaceType=flashlight

p. 91 Helmeted statue (upper left) https://www.pinterest.com/pin/maya-rise-and-fall-ancient-mayan-artifact-from-guatemala--309129961935337295/

p. 91-92 Lizard-like statue https://pagaian.org/2020/06/13/mother-and-child-her-eternal-cosmogenesis/

p. 92 big-eyed face: upper right; https://www.bushwalkingholidays.org/galleries/images/photos/isdell/wandjinna.jpg

p. 92 lower left two beings with radiant halos https://runews24.ru/science/07/11/2018/f812e420f68fc38900592d19f37d95d0

p. 93 creature with crown aboriginal art: https://en.wikipedia.org/wiki/X-ray_style_art#/media/File:Aboriginal_Art_Australia(3).jpg

p. 93 Large headed creatures in a row; source: trust for African art

p. 94 Egyptian tailed creature talking, temple of Hathor source: https://4.bp.blogspot.com/-CHAWZstmpko/WFjhGCC_JFI/AAAAAAAAD4A/yG1Bt2x52CUbLJGTeqQWl2s5Saq6AMPcwCEw/s1600/Dendera%2Bmonkeys.jpg

p. 95 Utah rock art: https://gearjunkie.com/outdoor/hiking/hike-camp-dinosaur-national-monument-guide

p. 95 Reptilian figures, Wikipedia, Vinca culture

p. 96 Wild horse canyon Wikipedia

p. 96 Sego Canyon https://www.greatamericanhikes.com/post/the-haunting-rock-art-of-sego-canyon

p. 98 Panel of rock art photos: PD, Vantage Washington, http://users.stlcc.edu/mfuller/vantageRock.html

Ginkgo Petrified Forest, https://tripwow.tripadvisor.com/slideshow-photo/ginkgo-petrified-forest-state-park-by-travelpod-member-drfumblefinger-vantage-united-states.html?sid=11520682&fid=tp-3

Burrup Rock Australia, http://www.crystalinks.com/petroglyphs1.htmlInner http://www.australiasnorthwest.com/Destinations/The_Pilbara/the-burrup-peninsula

Mongolia, http://www.chinasage.info/dynastyearly.htm and Tassil Niger. http://www.rockartscandinavia.com/images/articles/coulsona10.pdf

p. 98 African rock art https://brunoclaessens.com/2013/08/african-rock-art/

p. 98 Africa rock art https://www.pinterest.com/pin/aboriginal-cave-painting-of-the-lightning-brothers--346073552628366349/

p. 99 Painting "A summer's triumph" source: http://www.ancient-origins.net/unexplained-phenomena/unidentified-flying-objects-sky-summer-s-tapestry-001118

p. 100 Kosovo fresco; PD source: http://www.blastr.com/2011-12-7/12-ancient-paintings-containing-surprising-evidence-aliens

p. 105 Chinese and Mexican pyramids

p. 106 Giza pyramid plane PD

p. 107 Aerial view of Giza pyramid plane and Orion belt

p. 109 Orion belt & aerial view of Egyptian and Chinese pyramid planes

p. 109 China, Mexico, Egyptian pyramids & Orion: PD http://www.dzr.org.ro/ce-se-ascunde-si-ce-este-cu-adevarat-in-interiorul-unei-piramide-egiptene-imagini-realizate-recent-cu-o-tehnologie-de-ultima-ora/

p. 111 Thornborough henge and schematic http://www.ancient-origins.net/ancient-places-europe/sacred-prehistoric-neolithic-complex-thornborough-henges-003652

p. 112 Thornborough henge with Orion belt superimposed http://www.ancient-origins.net/ancient-places-europe/sacred-prehistoric-neolithic-complex-thornborough-henges-003652

p. 112 Thornborough schematic

p. 113 Author at Teotihuacan

p. 116 Horus & Osiris

p. 118 Nefertiti with child PD

p 118 Mummified heads of Nefertiti, Tutankhamen & Ramses II; source http://www.ancient-wisdom.com/cranialdeformation.htm

p. 119 Peruvian elongated skulls

p. 121 Elongated skulls Malta source http://nexusilluminati.blogspot.mx/2013/03/mysterious-elongated-skulls-throughout.html

p. 123 Pope Benedict, Catholic cardinals & Akhenaten (PD) http://www.rte.ie/news/2016/0909/815398-former-pope-benedict/Catholic cardinals: http://www.asiantribune.com/node/61938

p. 125 Dogon engravings Dogon engraving; PD http://ancientufo.org/2014/08/the-unexplained-knowledge-of-malis-dogon-tribe/

p. 133 Sumerian clay tablet

p. 135 Author photo

p. 139 footprint in volcanic ash. Sylvia Gonzalez, "Ancient Human Footprints Found In Mexico, Mason Inman, July 2005, *Science*, https://www.science.org/content/article/ancient-human-footprints-found-mexico

p. 140 Voynich PD https://en.wikipedia.org/wiki/Voynich_manuscript

p. 142 Milky Way and Voynich drawings; http //en.wikipedia.org/wiki/Milky_Way

p. 144 Sego Canyon pictograph wrote for permission. source: http://albertde-bruijn.com/home/archives/2842

p. 150 Kirlian photo PD https://en.wikipedia.org/wiki/Kirlian_photography

p. 153 Neanderthal PD Wikipedia

p. 155 Denisovan. https://www.nationalgeographic.com/science/2019/04/enigmatic-human-relative-outlived-neanderthals/

p. 157 Omo skull PD: source: http://www.templestudy.com/2012/09/

p. 158, 160 Mungo Man; "'Keeping place' plan to save Mungo Man for posterity." The Australian, theaustralian.com.au

p 164 Jerry Kroth

p. 167 Human brain. PD; https://www.health.qld.gov.au/__data/assets/image/0037/447787/frontal.gif

p 169 Late Bloomer figure

p. 171 Author at Great Pyramid

p. 177 Australian rock art. PD source: http://springtimeofnations.blogspot.mx/2012/06/mystery-airstrikes-in-azawad-historic.html

p. 180 Map of ice age at its maximum. PD: source: http://theresilientearth.com/?q=content/sun-not-co2-caused-ice-age-glaciers-melt

p. 181 Evolution of man graphic

p. 184 Pyramid of Giza and Pyramid of Pepi PD

p. 186 Great Pyramid PD

p. 187 Sumerian cuneiform and Egyptian hieroglyph PD

p. 189 Dixon relics PD https://theorkneynews.scot/2020/12/16/hidden-in-plain-sight-ancient-egyptian-relics-in-a-cigar-tin/

p. 194 Sumerian god Ashur PD

p. 195 Nativity scenes: http://more-sky.com/group/nativity-scene-desktop-wallpaper/

p. 197 Sphinx, http://www.guardians.net/egypt/sphinx/

p. Jerry Kroth

p. 203-204 Caroline natives and bamboo aircraft; PD source: http://www.janesoceania.com/oceania_cult/index.htm

p. 204 Nazca lines from space PD